Decentralized Finance (DeFi)

By

NFT Trending Crypto Art

MASTERSHIP BOOKS

UK | USA | Canada | Ireland | Australia
India | New Zealand | South Africa | China

Mastership Books is part of the United Arts Publishing House group of companies
based in London, England, UK.

First published by Mastership Books (London, UK), 2021

I S B N : 9 7 8 1 9 1 5 0 0 2 0 4 4

Cover design by Rich © United Arts Publishing (UK)
Text and internal design by Rich © United Arts Publishing (UK)
Image credits reserved.
Colour separation by Spitting Image Design Studio
Printed and bound in Great Britain

National Publications Association of Britain
London, England, United Kingdom.
Paper design UAP

ISBN: 978-1-915002-04-4 (paperback)

A723.5

Title: Decentralized Finance (DeFi)

Design, Bound & Printed:
London, England,
Great Britain.

Join Our

NFT Crypto Art & DEFI

Entrepreneur

Power Group

To help reinforce the learning's from our books, I strongly suggest you join our well-informed powerhouse community on Facebook.

Here, you will connect and share with other like-minded people to support your journey and help you grow.

>>>Click here to join Our Personal Growth Support Group <<<

News Site Here:

https://www.facebook.com/NFT-Trending-104426528387351

Community Group:

https://www.facebook.com/groups/nfttrending/

Want Free Goodies?

Email us at:

mindsetmastership@gmail.com

Follow us on Instagram!

@MindsetMastership

CONTENTS

0

INTRODUCTION TO
DECENTRALIZED FINANCE

Decentralized finance is a term used to describe a group of blockchain protocols that aim to give open access to conventional financial services. The apps that make up the fast-growing DeFi industry are built on permissionless platforms such as Ethereum, allowing users to participate in a range of common financial transactions without relying on a trusted central authority.

The idea of DeFi began with the introduction of Bitcoin in 2009. Bitcoin was the first feasible system to allow value transmission without relying on trusted intermediaries, despite not being the first attempt to establish a completely digital currency. Bitcoin was the first to present a blockchain network to the public, but its primary focus on security implies that the platform isn't adaptable. Suggested improvements take a long time to execute, even when they are broadly accepted.

Several more important DeFi protocols became operational in 2018 and 2019. Nevertheless, their use was restricted until the year that followed. The overall sum locked in all DeFi systems reached $1 billion for the first time in the first quarter of 2020 — nonetheless, the fledgling industry experienced severe losses. However, in the summer of the same year, DeFi came back stronger than ever. DeFi Pulse, a data and analytics business, said that the entire value locked in DeFi was about $7.8 billion by August 2020 1. It had reached an all-time high of $9.54 billion by

September 2.

Decentralized finance is one of crypto's fastest-developing and most exciting niches. Even though the industry is still in its early stages, significant DeFi networks have already received more than $50 billion in funding. This trend is expected to continue; especially as Ethereum transaction prices push developers and consumers to other blockchain technology like Solana, EOS, Polkadot, etc.

Most of the conventional banking industry's key functions are already covered in DeFi. Decentralized lending, borrowing, trading have all garnered substantial user bases. DeFi's transformative potential makes it an industry ripe for growth. With this guide, you should be better able to avoid the hazards and understand more about the newest advancements in this exciting and developing sub-industry. Let's get started

DeFi Characteristics

So, let's take a look at some of DeFi's most significant characteristics by breaking it down into its components.

DeFi's Components

Comprehending the components of decentralized finance is the first and most important step in understanding its features. Decentralized finance components are essentially the same as those found in other financial networks already in use.

Smart contracts provide the basis for developing DeFi applications. A DeFi platform's components are key elements of DeFi. Each component serves a distinct purpose in developing the DeFi network and is an important element of DeFi. The four key layers that constitute the DeFi architecture and are critical to DeFi's popularity are listed below.

- Settlement Layer

The settlement layer is one of the most noteworthy DeFi components since it is the foundation for all other DeFi solutions. It features a public blockchain together with the native digital currency. The native digital currency is used in most decentralized finance app activities, which may or may not be exchanged in other marketplaces.

- Protocols Layer

Network protocols are rules and standards that are designed to regulate certain actions or processes. The protocol layer contains standards and regulations that all players must follow in a certain industry. Decentralized finance protocols provide interoperability, allowing many entities to collaborate on creating a service or software at the same time. In the Decentralized finance ecosystem, the protocol layer is critical for achieving the necessary liquidity levels.

- Application Layer

The application layer is one of decentralized finance's most noteworthy characteristics, and it readily explains the question, "Why is DeFi so famous?" The application layer, as implied by its name, is where user-facing programs are stored. As basic user-oriented services, the decentralized apps abstractly reflect the underlying protocols. This layer is home to several cryptocurrency apps, including loan services and decentralized exchanges.

- Aggregation Layer

One of the key characteristics of DeFi is the final layer in the DeFi technology stack. Aggregators integrate diverse apps from the previous layer to provide service to users in the aggregate layer. Aggregators, for instance, might aid in the smooth transfer of funds between different financial instruments, therefore increasing returns.

DeFi's Key Characteristics

It's also easy to gain a good sense of DeFi's features by considering the advantages it provides. The benefits demonstrate DeFi's value proposition or the capabilities that you may incorporate in the DeFi solutions. Here are some of the significant things that DeFi has to offer.

- **Permissionless**

The permissionless feature of decentralized finance apps is one of the most notable aspects of the technology. Decentralized finance does not follow the typical access norms of traditional banking. However, it adheres to the open, permissionless access approach. Anyone with a connection to the Internet and a cryptocurrency wallet might use DeFi solutions. You can use DeFi no matter where you are or how much money you have if you have these two requirements. As a result, DeFi may potentially let nearly anyone into the financial sector.

- Programmability

Programmability is another noteworthy characteristic of DeFI. It's important to note that the bulk of current DeFi solutions are built on the Ethereum network. As a result, DeFi's capacity to access smart contracts with increased possibilities of programmability may aid in automated operation. DeFi's programmability, on the other hand, offers up new possibilities for developing new financial and digital products. As a result, DeFi has all of the needed resources to handle any sort of traditional financial service activity.

- Transparency

Transparency would be one of the most noticeable and evident DeFi characteristics. In the Ethereum network, all actions are broadcasted to other participants on the network. All participants should double-check the transaction broadcasted to them. It's worth noting that all Ethereum addresses are essentially encoded keys with the added benefit of false anonymity.

The increased openness of transaction data allows for in-depth data analysis. Transparency, on the other hand, guarantees that every user has access to information regarding network activities. Ethereum and decentralized finance protocols based on Ethereum are likewise built with open access codes that can be viewed, audited, and developed by anybody.

- Immutability

One of the potential solutions to the question of "why is DeFi so famous?" is the notion of financial inclusion for all. But with worries about immutability, data integrity is required for the interchange of information and financial activities in decentralized finance. As a result, tamper-proof data management is critical across the network's decentralized design.

As a result, it has the potential to increase security and audit scope significantly. Immutability is a necessary characteristic and a significant value benefit of integrating blockchain into the financial sector. Decentralized finance might guarantee integrity for all activities by ensuring safe and secure data transport without any illegal changes.

- Interoperability

DeFi's qualities also lead to interoperability as one of the most important criteria in today's modern financial services industry. Ethereum's modular software stack aids in guaranteeing that decentralized finance protocols and apps are designed to work together and support one another. Developers and product teams benefit from DeFi's versatility.

Developers may now quickly design new solutions or add new features to current protocols. Similarly, DeFi's characteristics may be used by developers and product teams to customize user interfaces and integrate third-party apps. One of the reasons decentralized finance protocols are regarded as legos is because of their interoperability. You must discover the proper technique to put two DeFi protocols together for certain use cases, much like you would with Lego bricks.

- Non-Custodial

The last and most significant characteristic of DeFi is that consumers retain total control over their assets and personal info. Web3 wallets like Metamask make it easier for participants to communicate with permissionless financial protocols and apps. DeFi solutions can help usher in a new era of customer-centric financial activities by giving you more control over your personal information.

Almost all discussions on the development and progress of DeFi include references to the various DeFi characteristics. DeFi protocols include a variety of characteristics that make them ideal for a variety of DeFi application cases, including:

- Asset management
- Decentralized Autonomous Organizations (DAOs)
- Borrowing and lending
- Gaming
- Insurance
- Decentralized Exchanges
- Data and analytics
- Margin trading
- Staking
- Tokenization

DeFi Products

The objective of DeFi product development is to remove the intermediaries in banking services, including loans, trading, investing, payment, insurance, and others.

DeFi products are modular due to the way they are constructed, which means that applications and protocols may be added to and integrated. In addition to the advantages of the blockchain network, this gives you a lot more freedom and diversity in your services.

Borrowing

There are two types of borrowing money in decentralized finance.

- Peer-to-peer

A borrower will borrow straight from a particular lender.

- Pool-based

Pool-based lending is when lenders contribute cash (liquidity) to a pool from which borrowers can borrow.

Using a decentralized lender has several advantages. These include:

- Borrowing Privacy

Today, everything about lending and borrowing money focuses on the people involved. Before lending, traditional financial institutions have to determine if you're going to pay back the loan.

Decentralized finance operates without requiring either side to reveal their identity. Instead, the borrower must put up collateral, which is immediately transferred to the lender if the payment is not made. NFTs are even accepted as security by some lenders. NFTs are a deed to a one-of-a-kind item, such as a painting.

This enables you to borrow money without having to provide any personal details or undergo a credit check.

- Access to Global Funding

You have access to funds placed from all over the world in decentralized finance, not simply the cash held by your selected bank or organization. This increases credit availability and lowers interest rates.

- Tax-efficiencies

Borrowing can let you get the money you need without having to sell any of your ETH (a taxable event). ETH can be used as security for a stablecoin loan instead. This provides you with the necessary cash flow while allowing you to maintain your ETH. Stablecoins are tokens that don't vary in value like ETH and are thus preferable for when you need cash.

- Flash Loans

Flash loans are decentralized lending that allows you to borrow money without putting up any collateral or giving any personal details.

They are now inaccessible to non-technical people, but they indicate what may be the case in the future for everyone. It operates on the principle that the loan is obtained and repaid in the same transaction. If it is not repaid, the transaction is treated as if it never occurred.

Liquidity pools hold the money that is often utilized (big pools of funds used for borrowing). If they aren't utilized at the time, someone can borrow them, trade with them, and repay them in full at the same time they were borrowed. This necessitates a great deal of reasoning in a highly customized transaction. A basic example is someone utilizing a flash loan to borrow as much of an asset at one price to sell it on a different market at a higher price.

As a result, in a single transaction, the following occurs:

You borrow X amount of $asset from exchange A for $1.00.

You sell X $asset for $1.10 on exchange B.

You repay the loan used to exchange A.

You keep the profit once the transaction charge is deducted.

The transaction will completely fail if exchange B's supply fell unexpectedly and the user cannot acquire enough to pay the original debt.

You'd need a huge sum of money to accomplish the preceding scenario in the conventional banking industry. These money-making techniques are only available to people who already have a lot of money. Flash loans illustrate a future in which having money isn't always required to make money. A chapter of this book is dedicated to flash loans.

Lending

DeFi enables anybody to take out or provide a loan without the need for third-party authorization. The great bulk of lending products over-collateralize outstanding loans with popular cryptos like Ether ($ETH).

Maintenance allowances and interest rates may now be written directly into a borrowing contract, with liquidations occurring immediately if an account balance falls short of a set collateral ratio, thanks to the introduction of smart contracts.

The interest generated by providing various crypto differs depending on the asset and platform employed. By lending your cryptocurrencies, you may earn interest and see your cash increase in real-time. Presently, interest rates are far greater than those offered by your traditional bank (assuming you're lucky enough to have one). Here's an illustration:

You lend your stablecoin, 100 Dai, to a service like Aave.

You will be given 100 Aave Dai (aDai), a token that symbolizes the Dai you have borrowed.

Your aDai will rise in line with interest rates, and you'll notice your wallet balance grow. After some days or perhaps hours, depending on the APR, your wallet balance will reflect something like 100.1234!

At any moment, you can withdraw an amount of ordinary Dai equivalent to your aDai balance.

No-loss lotteries

No-loss Lotteries, such as PoolTogether, are a fun and inventive new method to save money.

You spend 100 Dai tokens on 100 tickets.

Your 100 tickets are represented by 100 plDai.

When one of your tickets is chosen as the winner, the prize pool will be deducted from your plDai balance.

If you don't win, the 100 plDai will be carried over to the next week's draw.

At any moment, you can withdraw an amount of ordinary Dai equivalent to the plDai balance.

Fund Your Ideas

Ethereum is an excellent crowdfunding platform:

Interested funders may come from any place – Anyone around the globe may use Ethereum and its tokens. It's open, which allows fundraisers to see how much money they've raised. You may even track how money is used after they've been allocated. For instance, fundraisers could set up automatic refunds if a certain deadline or minimum amount isn't fulfilled.

Advanced Trading

There are more sophisticated choices for traders that like a bit more control. Limit orders, perpetual, margin trading, and other strategies are all viable options. You have access to worldwide liquidity with decentralized trading, the marketplace seldom closes, and you have control over the assets.

If you're using a centralized exchange, you must first deposit your funds and trust them to keep them safe. Your funds are in danger while they are deposited because centralized exchanges are appealing prey for fraudsters.

Decentralized Exchanges (DEXs)

DEXs are decentralized exchanges that enable participants to switch assets without transferring ownership of the underlying collateral. DEXs seek to enable secure, transparent trading across a variety of trade pairings. DEXs have witnessed huge advances in usage recently, with some even giving liquidity incentives for money contributed to the platform.

Derivatives

In the conventional banking industry, a derivative is a contract whose value is generated from an arrangement based on the performance of an underlying asset. Futures, forwards, options, and swaps are the four basic forms of derivative agreements. Users profit from open, interoperable, and programmable derivative agreement payments thanks to the introduction of DeFi. Smart contracts can create tokenized derivatives that give access to both ends of an asset's performance.

Wallets

Wallets are an important interface for dealing with DeFi products. While their fundamental product and asset backing may differ, the rising DeFi narrative has resulted in dramatic gains in usage and access across the board.

Asset Management

With so many DeFi solutions on the market, it's critical to have tools in place to track and manage assets effectively. In keeping with the permissionless characteristics of the DeFi network, these asset management initiatives allow participants to monitor their balances across multiple tokens, assets, and services in an easy manner.

Insurance

Participants may take out policies on smart contracts, money, or any other cryptocurrency using pooled funds and reserves thanks to decentralized insurance standards. Presently, this area is relatively small, but we expect it to play a significant part in the system's overall development.

Savings

There are a few DeFi projects that provide new and creative methods to profit from cryptocurrency savings. Because there is no borrower on the opposite side of the table, this is not the same as lending.

DeFi's Fundamental Objective

The primary objective of DeFi tokens and DeFi, in general, is to create a transparent and open access financial services network. The many theoretical and practical views on DeFi suggest that DeFi is accessible to everybody. DeFi, however, also allows for the absence of any centralized power interference.

DeFi's Advantages

DeFi's advantages also highlight the underlying causes that drive interest in decentralized finance tokens. Many of these advantages have been covered earlier. However, here are some of the most important benefits of DeFi.

- Accessibility

DeFi has the potential to revolutionize financial inclusion for the world's 1.7 billion individuals who do not have access to financial services.

- Trustless

Participants do not have to place their trust in DeFi apps. Participants do not have to put their confidence in any authority to manage their assets and investments as they have complete control over them.

- Low-Cost

DeFi apps do not have a significant operational and maintenance expense. DeFi apps may also settle disputes more effectively since they don't need to rely on intermediaries. This can save costs on call centers operatives and administration staff.

DeFi Tokens

Now that you've learned about the many benefits of DeFi and their importance, you might be wondering what DeFi tokens are. The tokens may be thought of as decentralized financial apps that run on blockchain systems and replicate key ideas inherent in the conventional finance and banking industry.

Many token distribution schemes have evolved throughout the years, with Simple Agreement for Future Governance being the most recent example. Because DeFi tokens are linked to monetary worth, you don't have to be concerned about their price. Tokens having a specified monetary value will be available on DeFi platforms that offer exchange and lending services. Users are rewarded for utilizing the platform's native currency with a lower interest rate or earning free tokens for completing specific tasks.

Popular DeFi Tokens

The emergence of many DeFi platforms has undoubtedly increased search queries such as "what are the best DeFi tokens?" with many promising responses. Many DeFi platforms indicate a large number of DeFi tokens and the challenge in selecting the best one. So, let's take a look at some of the most popular tokens in the DeFi market right now.

- MKR

MKR is the DeFi token of MakerDAO, a famous DeFi program. MakerDAO provides a diverse range of services to consumers under several identities. With its subsidiary Oasis and its stablecoin, Dai, the firm provides decentralized lending services. Holders of MKR tokens are in charge of the Maker Protocol's governance. Modifications to Dai stablecoin regulations, governance improvements, and the choice of additional collateral types are all part of the protocol.

- COMP

Comp is Compound's DeFi token, which is a prominent decentralized cryptocurrency lending platform. Comp is unquestionably one of the greatest lending and borrowing assets in the decentralized finance cryptocurrency ecosystem. The supply and demand of cryptocurrencies determine interest rates, and the amount of interest determines how much Comp is allocated to markets. Comp is also crucial for the administration of key protocol-related decisions.

- AAVE

When it comes to the question of "what are the finest DeFi tokens?" among other options, the word Aave comes up frequently. Aave is a DeFi lending platform that also has its native token, LEND. The native coin assists holders in getting lower fees, with intentions to use it as a governance stake in the future. It might also act as a one-of-a-kind initial line of defense for any outstanding debts.

- ALGOR

The ALGO DeFi token is the native currency of Algorand, a decentralized application development platform based on Ethereum. Algorand is well-suited to accepting loans and allowing decentralized trade and a variety of other applications. ALGO is well suited to compensating network members with network verification privileges.

- ZRX

ZRX is another famous DeFi token to keep an eye out for. The permissionless liquidity system's native token is 0x. Along with building new decentralized exchanges, 0x may bring liquidity providers to one page. The DeFi token ZRX may be used for a wide range of applications. Staking ZRX with market makers on 0x, for instance, might help you earn incentives. In addition, ZRX is well-suited to governance applications.

1

DEFI LENDING PLATFORMS

DeFi Lending and Borrowing

DeFi Lending

DeFi lending, which involves a user depositing funds into a protocol, is similar to a typical fund deposit or investment that pays interest over time. Lenders get not just interest on their digital funds but also a governance token or DAI as a bonus: Compound COMP, Aave produces LEND, and Maker offers DAI. The 3-5 percent interest rate for lending is better than many banks for ordinary customers, but it may not be sufficient to offset the ever-present danger of smart contract abuses. On the other hand, these rates look to be highly appealing to high-capital investors, hedge funds, or institutions, especially when applied to stablecoins like USDT, USDC, or DAI. Lending may also assist in minimizing the risks of market fluctuations by allowing users to make money without having to trade.

For the most part, lending rates vary among each Ethereum block. Price oracles assist in determining the optimal annual percentage yield (APY), which varies to keep the procedure functioning smoothly. Users who lend bitcoin are rewarded with platform-specific tokens. For instance, if you put 1 ETH on Compound, you will receive 50 cETH tokens. Such tokens are used by the platforms to calculate the accumulated interest and are required to withdraw your funds.

DeFi Borrowing

The majority of funds on a lending platform aren't there just to earn interest. Becoming a lender is just a small part of the equation; the true magic occurs when considering the range of options available to lenders. However, first and foremost, it's critical to comprehend collateral.

The usage of decentralized technologies does not need authorization. As a result, traditional assessments such as credit score, equity, or income cannot be used to establish a safe loan amount. Lending sites, on the other hand, demand borrowers to provide cryptocurrency assets as collateral. Over-collateralization is a common feature of DeFi loans. This implies that participants can only collect a part of the collateral they have provided: When you lend $10,000 in ETH, you could get up to $7,500 in DAI or other assets (roughly 75 percent of your collateral). This may sound paradoxical at first, but it's important to verify that all users can repay their loan; the collateralized assets may be liquidated when you can't repay the loan.

Compound exclusively provides variable interest rates for loans, whereas Aave users may choose between fixed and variable interest rates. Variable rates subject debtors to liquidation if the annual percentage yield (APY) surpasses a specific level. These variable-rate loans need frequent focus and dedication. Nevertheless, based on the present amount lent and borrowed, they are usually less expensive than fixed-rate loans.

DeFi Lending and Borrowing

Why would you like to borrow against your assets for a loan, which is substantially below your collateral? Over-collateralization poses a significant concern. This is because many cryptocurrency investors do not like to sell their most valuable assets. They may free up liquidity without having to trade by lending their money. When someone owns $50,000 in ETH and doesn't wish to sell it, they can give it to a lending protocol and borrow up to 75% of its worth.

This opens up a whole new universe of opportunities for cryptocurrency traders: they may perform open market margin trading, buy a coin they don't own for liquidity mining, or take out a quick loan for crises. All of this was accomplished without selling an asset. Hedge funds and organizations that hold cryptocurrency as part of their portfolio may find cryptocurrency lending especially beneficial. They may take out a loan against their cryptocurrencies and convert them to regular financial instruments. These are only a handful of the numerous applications.

Of course, both financing methods are risky. DeFi, on the other hand, has established a modern system for borrowers and lenders, one that would not be feasible without the use of cryptos and blockchain technology.

DeFi Lending Market

Compound, Aave, and Maker are the key platforms in DeFi lending. According to DefiPulse, compound controls over 50% of the loan market and more than 60% borrowing.

Participants that provide crypto to Compound earn cTokens, which are ERC20 tokens that may be claimed at any moment for the asset class. Like most of its rivals, Compound uses a system to generate interest rates for the cryptos it supports, including DAI, ETH, WBTC, USDC, USDT, UNI, ZRX, BAT SAI, and REP. Anybody with an Ethereum wallet may use Compound to deposit cryptocurrency and receive interest straightaway or borrow against collateral.

All debts are over collateralized when it comes to borrowing. For example, if you plan to borrow DAI, you'll have to over-collateralize the loan with another asset like WBTC.

According to Messari's Ryan Selkis, Compound's rival Maker may have been the crucial component element that created the basis for DeFi's credit markets, stablecoin markets, and eventually, 2020's DeFi bull run. Individuals could get loans in the Maker community that is priced in the system's dollar-pegged Dai stablecoin. Participants must keep a buffer of at least 150 percent over-collateralization of the amount of Dai they have coined to guarantee system stability.

Another prominent lending protocol is Lendefi. It is remarkable because it allows participants to access undercollateralized loans, allowing them to get loan values that might otherwise be unavailable. Lendefi does this by utilizing decentralized liquidity pools like Uniswap that enable P2P trading. The network manages the protocol through a Decentralized Autonomous Organization (DAO), allowing users to participate in the platform's direction actively.

Essentially, the conditions of cryptocurrency lending are determined by the asset being deposited or borrowed, as well as the payback period chosen. While DefiPulse gives a glimpse of current rates, LoanScan covers a larger range of lenders, including Nexo and BlockFi and Yearn, dYdX, and Curve. In general, interest rates start at approximately 0.03 percent and can reach above 50% APY in some circumstances.

Risks Associated with Lending

If you believe DeFi lending is a land of milk and honey, you're mistaken. While these protocols are generally secure and protected, attackers have taken advantage of vulnerabilities and stolen assets from users. In a sophisticated cyberattack last year, the DeFi lending market lost about $8 million worth of DAI stablecoins. In another attack, DeFi platform, Akropolis, lost $2 million.

The fact is that assets held in DeFi networks are only as safe as the programming that underpins them. That's not to mention the dangers associated with market fluctuations, which may lead participants to become undercollateralized, requiring them to contribute more funds to prevent liquidation.

In the end, DeFi lending is still in its infancy. While these procedures provide borrowers and lenders with a gold-plated way to avoid banks, it is essential to do your homework and put your trust in a trustworthy, safe network that provides a sense of peace. Cryptocurrency lending may be a profitable strategy to release liquidity and create passive income from your assets if you do your homework correctly.

Lending Platforms

Source of Interest

Before we analyze common lending platforms and go into the most important factors to take into account when selecting a lending platform, it's critical first to comprehend where your interest will come from. The borrowers pay the interest.

Borrowing backed assets is also possible in every lending platform, with borrowers repaying their loans with interest. The lenders get this returned interstate money. Because there are typically more lenders (liquidity suppliers) than borrowers, borrowing interest rates are greater.

Selecting a DeFi Lending Platform

When we plan to select a DeFi lending platform, there is no one-size-fits-all solution. Popular platforms such as AAVE, Compound, and dYdX all operate similarly. Thus it all boils down to risk, reward, and personal preferences. With that in mind, here are a few key factors to keep in mind.

APR (Annual Percentage Rate) Comparison

When considering lending platforms, the projected yield is often always the first item that comes up. In other words, users are curious about the platform's APR (Annual Percentage Rate) for deposited monies.

If nothing else happens, the APR decides how much money may be anticipated in a year. The issue is that the APR used in DeFi platforms is extremely volatile. The ratio between the monies provided to the pool and the amounts borrowed (the pool's usage) determines the APR. When a large portion of the pool is borrowed, the interest rate rises, encouraging lenders to provide liquidity to the pool while discouraging borrowers from taking high-interest loans.

This lender-to-borrower ratio might fluctuate dramatically from day to day and even block to block. Because of the advent of the Compound COMP governance token, the APR of the BAT token on the BAT platform jumped from 0% to 27% and back to 0% in a short period.

Cost of Transactions and Gas

As you may be aware, every transaction on Ethereum incurs a fee that is paid to miners in exchange for the platform's security. Three key factors determine every transaction's USD cost:

Gas Amount: The intricacy of the calculation is measured by the amount of gas used. Every transaction requires more gas as the smart contract becomes more sophisticated.

Price of Gas: The greater the need for transactions, the greater the price. Because the amount of space available in each block is small, adding a transaction will increase the cost.

Price of ETH: Gas is paid in ETH. Transactions get more costly as the value of ETH rises.

Transaction cost = GasAmount * GasPrice * EthPrice

The platform has nothing to do with the price of gas or ETH. The need for ETH and transactions on the Ethereum network decide them. The quantity of gas needed in using a platform depends on protocol, token, and whether the transaction is a deposit or withdrawal from the protocol. The transaction cost is calculated by multiplying the amount by the gas and ETH prices, provided the gas needed.

Calculating the Figures

When determining whether or not lending is the best option for you, you must do the arithmetic. Making the best selection is dependent on the present APR of every protocol, present market prices, the length of time money is locked up, and if a minimum deposit amount is necessary.

Gas costs are quite unpredictable, and ETH and APR fluctuate regularly. On the bright side, gas prices are largely unrelated to the amount borrowed. Whether you deposit $100 or $100,000, the cost is nearly the same.

Security

When it comes to picking a protocol, security is crucial. The fact that all DeFi protocols operate on a blockchain network does not guarantee that they are secure. After all, the smart contracts that enable these protocols are only programs, and every software program might contain defects that result in a loss of cash and the protocol's entire breakdown.

The bZx Fulcrum protocol, which recently lost $350K in user cash owing to a contract issue, is a warning example with important lessons. No deposits or withdrawals were permitted while the platform was momentarily stopped. (As a side point, Compound's CEO chastised the bZx platform for plagiarizing the Compound software without fully comprehending the consequences.)

It's essential to determine whether a lending platform is well-known and trustworthy, just as it is with any other financial service. AAVE, Compound, and dYdX were all audited by reputable companies. Most of them were subjected to many audits. While an audit isn't an assurance of security because code might still be flawed, the absence of one should raise concerns about the protocol's development.

Other important security considerations include whether the contracts are controlled by a single organization, which owns the keys, and if the business promises a return if money is stolen. These characteristics might be difficult to determine, but that should not deter users from undertaking comprehensive research. This is especially true for people who are putting down huge sums of money.

Supported Assets

Not every network accepts the same set of tokens. Only USDC, DAI, and ETH are supported by dYdX. AAVE and Compound both support a significantly larger number of tokens. Lending the UNI token, for instance, is presently only accessible in Compound. Likewise, AAVE is the sole choice for KNC.

Many popular DeFi tokens, such as the above-mentioned AAVE, UNI (also accessible for purchase and trade), and KNC tokens, have been included in ZenGo's supported assets list. ZenGo Savings supports the Compound protocol. All of Compound's assets may be used to save money simply and safely, with the bonus of generating COMP tokens.

Now let's turn our attention to some of these platforms

Aave

Aave supports a variety of cryptos, including stablecoins and altcoins. Individuals may borrow these cryptocurrencies at fixed or variable interest rates, or they might lend them to liquidity pools and receive interest on deposits. AAVE is the native token used by the Aave lending system. The AAVE token is a governance token that may be staked in exchange for fees and other benefits.

Aave is a non-custodial service, which means that cryptocurrencies are never stored by Aave directly. The participant has complete control over their deposits.

How to Lend on Aave

Aave allows participants to deposit a variety of cryptos in exchange for variable rates of interest. Certain cryptos have greater yields than others, but interest rates are always fluctuating. To deposit Aave, you'll have to log in with a web 3.0 Ethereum digital wallet like Metamask or Formatic:

Select "deposit" next to the asset you'd like to lend on the Aave site. Via the web 3.0 wallet, input the amount, and complete the transaction. You will get the related aTokens in the digital wallet after the transaction is completed.

The aTokens will subsequently start earning interest.

aTokens

aTokens refers to Aave interest-bearing tokens. Deposits of crypto to Aave lead to the generation of aTokens, which are derivative tokens. The worth of such aTokens is equivalent to the value of the deposited crypto.

If a participant deposits 100 DAI on Aave, they will get 100 aDAI tokens. The 100 aDAI will generate interest and may be sold on any DeFi platform that allowed them or swapped for the initial asset on Aave.

How to Borrow on Aave

Participants may borrow at a variable or fixed interest rate from any of the crypto pools featured on Aave. Participants will require to have access to a web 3.0 digital wallet to achieve this:

Participants would need to deposit a digital asset to be applied as collateral before borrowing. The amount of money you can borrow will be proportional to the amount you put up as collateral. It's worth noting that all loans are over-collateralized, which means the deposit must be more significant than the loan.

After you've deposited your collateral, go to the Aave home page and look for the crypto you want to borrow. On the right side of the screen, tap "borrow."

After that, you'll be asked to link the web 3.0 digital wallet. Once you've established a connection, select the amount you'd want to borrow. After that, you have to decide whether the loan will have a fixed or variable interest rate. Select your desired interest rate and complete the transaction using the web 3.0 digital wallet.

Variable and Stable Interest Rates

On Aave, interest rates are continually changing and adapting based on Aave's liquidity pools. The is referred to as utilization rate.

Interest rates stay low when there is a large amount of cash in the liquidity pools, encouraging participants to take out loans. Whenever money is scarce, interest rates rise to reward lenders and motivate borrowers to repay their loans.

The demand in the Aave markets affects variable interest rates. Variable rates can help you save money on interest costs, but it all relies on market conditions. Variable rates are not appropriate for long-term financial planning due to their inconsistency.

Stable interest rates, on the other hand, give the borrower a more anticipated rate. It's important to note that fixed interest rates are not synonymous with traditional fixed rates. They are still able to move. They are, however, a far more stable type of variable interest rate that is less impacted by market fluctuation. Stable interest rates are frequently greater than variable interest rates because of the advantage of predictability.

Maker

Maker (MKR) is the Maker's primary utility and governance token. Maker is a decentralized autonomous organization (DAO) built on Ethereum, enabling anybody to lend and borrow crypto without a credit check. To do this, the platform combines sophisticated smart contracts with a specially pegged stablecoin.

The platform, for instance, was one of the first Ethereum-based tradeable tokens. Maker is now one of the most widely used Ethereum-based platforms. Maker CDP contracts have a total value of almost 2.1 million ETH.

How It Works

MKR is an important component of the Maker platform. MKR, for instance, may be used to transfer value internationally, similar to Bitcoin. This token may equally be used to pay transaction fees on the Maker platform. MKR may be transferred and received by any Ethereum account and any smart contract that has the MKR transfer function enabled.

DAI

MKR is meant to support DAI as a stablecoin. The MakerDAO uses CDP smart contracts to create DAI tokens. DAI was the first decentralized stablecoin on the Ethereum network. The Oasis Direct method, for example, is used to swap MKR, DAI, and ETH. MakerDAO's decentralized token trading platform is called Oasis Direct.

How to Buy Maker (MKR)

Maker (MKR) is traded on the following platforms.

- Kraken

For citizens of the United States, this is the greatest choice.

- Binance

Binance is the best cryptocurrency exchange for Australia, Canada, Singapore, the United Kingdom, and the rest of the world. MKR is not available for purchase on Binance in the United States.

- Easy Crypto

For Australia, New Zealand, and South Africa, this licensed exchange is the best option.

How to Store Maker

DAI and MKR may be saved in any ERC-20 supported wallet. Metamask is one of the prominent choices available currently. This wallet is available for free on Chrome and Brave, and it just takes few minutes to set up.

A hardware wallet is an ideal solution if you want to make a large investment in MRK or plan on Hodling this cryptocurrency for a long time. Hardware wallets protect your cryptocurrency in "cold storage" off the internet. This technique prevents internet attackers from gaining access to your assets. Maker is supported by both the Ledger Nano S and the more powerful Ledger Nano X.

Compound

Cryptocurrency traders use Compound to lend and borrow digital assets. Compound cryptocurrency is a blockchain-based decentralized application (dApp).

How It Works

The following assets are supported on Compound:

- Ether (ETH)
- Dai (DAI)
- Ox (ZRX)
- Tether (USDT)
- USD Coin (USDC)
- Wrapped BTC (WBTC)
- Sai (SAI)
- Augur (REP)
- Basic Attention Token (BAT)

cTokens

When participants deposit money on the Compound system's lending side, they are given cTokens, which are digital assets that reflect the amount placed. The Ethereum blockchain technology is used to create cTokens, which are ERC-20 tokens. On the Compound network, every cryptocurrency has its own cToken, such as cETH, cBAT, and cDAI. Participants get the token that corresponds to the cryptocurrency they deposited.

Holders of cTokens have full control over their public and private keys, exactly like Bitcoin or any crypto. In the end, the cToken may only be reclaimed for the cryptocurrency it reflects.

Interest Rate

Depending on the liquidity available for every crypto provided on the market, the Compound system estimates and provides interest rates constantly. The prices are continuously changing and fluctuate depending on market supply and demand. The interest rates are low when there are a lot of funds in the Compound wallet. Since there are so much funds accessible to borrowers, lenders don't make huge money in exchange for adding to the pool.

The interest rates are greater if the pool of money for particular crypto is small. This generates a persistent incentive for participants to invest in pools with less money to receive a higher return. Borrowers are also encouraged to borrow from big pools and refund their loans into smaller pools, resulting in reduced interest rates.

The Compound dashboard displays a yearly interest rate, which participants are given when they request a quotation. After every 15 seconds, any cTokens in a participant's possession grow by 1/2102400 of the current year's interest rate. The number of 15-second blocks in a year equals that proportion.

Lending

Cryptocurrency owners can lend any amount of their holdings, a process known as locking, sending, or depositing. This is like placing fiat cash into a savings account that instantly begins collecting interest. In contrast to depositing funds in a traditional bank account, the Compound dApp is decentralized, and the funds are placed in a big pool alongside other crypto traders' deposits. The currency that the lender deposits is the currency that they will be paid.

Borrowing

The capacity to borrow against deposited and locked money is another key element of the Compound system. Any participant contributing a portion of their crypto holdings to the Compound pool may borrow against those assets right away, with no credit check or other restrictions. The amount a person may borrow is determined by how much money they have on hand, and each crypto has its own set of rates.

To guarantee that their money is collateralized, borrowers have to deposit more than they expect to borrow. This indicates that money is available to repay the loan when the participant fails to pay the installments and interest. Because cryptocurrencies change in worth, if the collateralized amount drops in value, the borrower cToken smart contract shuts immediately whenever the value approaches the borrowed amount. The borrower maintains the cTokens they borrowed, but they forfeit the collateral they deposited if this happens.

Borrowers have to pay interest on the cash they borrow, much as they would if they borrowed from a traditional financial institution. The interest rates, which vary with every crypto on the site, are constantly determined and implemented by the Compound algorithm.

InstaDApp

InstaDApp is a DeFi platform that aims to provide a user-friendly interface for interacting with other DeFi protocols. Rather than needing four distinct apps to accomplish a single task, like lending and borrowing, InstaDApp offers a single platform that enables you to do everything in one spot.

How it Works

A participant must have an Ethereum web3 wallet such as MetaMask, Coinbase Wallet, or Trust Wallet to utilize InstaDApp and engage with its interface. They may then use the InstaDApp dashboard to handle all of their digital assets. All transactions on the site are conducted via smart contracts, with cryptocurrency assets held in a participant's contract wallet, ensuring that they have complete control over their assets. InstaDApp does not charge a fee; all a user has to do is make sure they have enough gas before starting one.

The four primary activities that you may engage in on the platform include:

- Lending - Deposit your cryptocurrency and get interest.
- Borrowing - Take out a loan right from the InstaDApp dashboard.
- Leverage - Increases the amount of money you can trade with.
- Swap - Exchange tokens from your web3 wallet in real-time.

Because this app is decentralized and employs smart contracts to incorporate other dApps, it may be used by anybody. Instadapp makes use of smart contracts, which transform basic user input into more sophisticated activities. The program then implements these contracts, providing users with a simple and intuitive method to manage their assets. What's even better about this platform is that there are no costs associated with it. Nevertheless, participants will not be free because they will still be responsible for paying for the gas required to complete the contracts.

Additionally, all of Instadapp's data is open to the public, and the platform does not store any of the participant's assets. This offers the participant the best of both worlds: the security benefits of decentralization combined with a centralized platform for asset management that offers a great user experience for participants.

Lending and Borrowing

By integrating into the DeFi protocol, Compound Finance, InstaDApp functions as an interface to facilitate lending and borrowing on its platform. InstaDApp's leverage and swap features are powered by Kyber Network, a technology that collects liquidity from a variety of sources. For instance, if a participant wants to make a leveraged transaction using Ether, InstaDApp interacts with Kyber to provide them with the extra Ether they need. Since Kyber has access to a larger selection of cryptos because of its liquidity reserves. An InstaDApp participant may swap their cryptocurrency for almost any other provided that Kyber supports it.

How to Get Started

Obtaining a web3 wallet is the first step to get started with Instadapp. MetaMask is a good example of a useful wallet since it is simple to set up and offers many features.

Connecting the Wallet

You'll have to link your wallet to Instadapp once you've created one. The first step is to hit the connect option at the top right corner of the dashboard. When you hit the button, you'll be given two options; one is to use WalletConnect to link your wallet, and the other is to use Coinbase Wallet to link the wallets. If you use the MetaMask browser extension, you will be given a third option that enables you to link MetaMask to Instadapp.

If you select one of the first two options, a QR code will display, which you must scan to link the wallet. The Coinbase Wallet app is required for the Coinbase option. This app is available on both the App Store and Google Play. You may link the wallet to Instadapp immediately after the app is installed by scanning the QR code supplied by Instadapp. The wallet will connect after you scan the QR code.

Create a User Account

The next thing is to select the "Create Account" option, which will send a push notification to the wallet application.

Nevertheless, you will be required to pay a miner fee to register the account and complete the setup procedure. The price for linking the wallet is about 0.0322 ETH, which is almost $53. This is a rather high cost. However, it is due to the present high gas fee for Ethereum.

Additionally, to link your wallet to Instadapp, you must have adequate cash in your account. The request will remain pending until you pay for the gas, at which point it will be executed as quickly as feasible. The timeframe may vary, and you may expedite the procedure by paying a greater price. Your Instadapp account will be active after the program verifies the transaction.

2

DECENTRALIZED EXCHANGES

Decentralized exchanges, also referred to as DEXs, are peer-to-peer markets where crypto investors conduct transactions without entrusting their assets to an intermediary. Such transactions are made possible via smart contracts, which are self-controlling contracts written in code.

DEXs were developed to eliminate the need for any authority to supervise and allow deals inside a given exchange. Peer-to-peer (P2P) cryptocurrency trading is possible on decentralized exchanges. Peer-to-peer refers to a cryptocurrency marketplace that connects buyers and sellers. They are often non-custodial, meaning that participants retain control over their wallet's private keys. Participants may access their cryptocurrency via a private key, which is a kind of sophisticated encryption. After authenticating into the decentralized exchange with their private key, participants may instantly see their cryptocurrency balances. They will not be needed to provide any personal details or addresses, which is ideal for those who value their privacy.

How DEXs Work

All trade induces a transaction cost in addition to the trading fee since DEXs are built on top of blockchain technology that allows smart contracts and where participants maintain custody of their assets. To use decentralized exchanges, traders engage with smart contracts on the blockchain network.

Automated market makers (AMM), Order books DEXs, and DEX aggregators are the primary forms of DEXs. All of them employ smart contracts to enable participants to trade freely with one another. The early DEXs utilized order books that were comparable to those used by centralized exchanges.

Automated Market Makers (AMMs)

An automated market maker (AMM) system based on smart contracts was developed to address the liquidity issue.

These AMMs use blockchain oracles, which are blockchain-supported services that offer data from exchanges and other networks to establish the price of traded assets. Rather than matching buy and sell orders, these DEXs' smart contracts employ liquidity pools, which are pre-funded pools of assets.

Other participants finance the pools, and they are then subject to the transaction fees charged by the system for implementing transactions on that pair. To earn income on their crypto assets, these liquidity providers must deposit an equal amount of all assets in the trading pair. A process referred to as liquidity mining. If they try to deposit more of an asset than another, the smart contract that runs the pool invalidates the transaction.

Traders can utilize liquidity pools to implement orders or get interest without needing permission or trust. Because the AMM concept has a disadvantage when there isn't sufficient liquidity, these exchanges are frequently rated based on the amount of cash held in their smart contracts, known as total value locked (TVL).

Order Book DEXs

Order books keep track of all open purchase and sell orders for certain asset pairs. Purchase orders show a trader's willingness to purchase or bid for an asset at a specified price, while sell orders show an investor's willingness to sell or ask for the item in question at a given price. The size of the order book and the market price on the exchange is determined by the gap between these values.

There are two forms of order book DEXs. They are on-chain order books and off-chain order books. Open order information is frequently held on-chain by DEXs that employ order books, while participants' money stays in their wallets. Investors on these exchanges could leverage their holdings by borrowing money from lenders on the site. Leveraged trading improves a trade's earning potential while equally increasing the danger of liquidation. It boosts the size of the position with borrowed money that must be returned even if the investors lose their stake.

On the other hand, DEX platforms that keep their order books off the blockchain simply execute orders on the blockchain to provide traders with the advantages of centralized exchanges. Exchanges can save money and time by using off-chain order books to ensure that deals are performed at the prices that participants want.

These exchanges equally enable participants to lend their cash to other investors to provide leveraged trading opportunities. Loaned money accrue interest with time and is protected by the exchange's liquidation process, which ensures that lenders are compensated even if investors lose theirs.

It's worth noting that order book DEXs frequently have liquidity difficulties. Traders generally adhere to centralized platforms as they are contending with centralized exchanges and pay extra expenses due to the fees needed to transact on-chain. While decentralized exchanges with off-chain order books decrease these expenses, the requirement to deposit cash in smart contracts introduces smart contract-related vulnerabilities.

DEX aggregators

To deal with liquidity issues, DEX aggregators employ a variety of protocols and methods. These platforms effectively pool liquidity from many decentralized exchanges to reduce slippage on big orders, reduce swap fees and token prices, and provide traders with a better price in the shortest amount of time.

DEX aggregators' other major aims are to protect participants from the price effect and reduce the chance of unsuccessful transactions. Certain DEX aggregators additionally leverage liquidity from centralized systems to give a better user experience, all while staying non-custodial through the usage of specialized centralized exchange integrations.

DYdX

Although dYdX is restricted basically as a trading platform, it is among the most sophisticated as an open, trustless, and non-custodial platform. Fundamental trading between three simple assets (ETH, DAI, and USDC), lending funds to earn interest, and major types of margin trading (isolated margin trading and cross margin trading) are presently available on the platform. These are basic instruments for a seasoned trader, but they represent a big step forward for the DeFi market.

dYdX closes deals on the blockchain network but employs an off-chain matching algorithm to provide the same implementation speeds as centralized exchanges while maintaining the security of non-custodial assets.

How it Works

Everybody engages in one "universal lending pool," rather than separate borrowers and lenders offering and receiving loan proposals. Every asset has its lending pool, which is handled by smart contracts, allowing for instant withdrawals, borrowing, and lending without necessarily waiting for matches or adequate money. The interest rates of any asset are determined by the interplay between borrowers and lenders– demand and supply.

Margin Trading

dYdX enables margin trading in conjunction with spot trading (changing one digital token to another), allowing cryptocurrency investors to perform more complex trading methods.

Margin trading, equally referred to as leveraged trading, enables traders to take greater holdings by borrowing money from a third party. Not only does this enable expert traders to increase their trading earnings (and losses!), but it equally allows regular traders to take greater holdings with a smaller amount of initial cash.

Example:

If you're optimistic about Ethereum, you might use a $1,000 initial margin to enter a 5x leveraged trade on ETH/USDC. In that situation, you'd have $5,000 worth of ETH/USDC exposure. If the price increased by 10%, the position would be increased by $500 (rather than just $100), giving you a 50% return on your investment if you closed it out at that time.

If the bullish conviction is incorrect and the price of ETH falls by $10 against USDC, the worth of the $1000 5x leveraged position will be reduced to $500. A 50% drop in value.

All you have to do to begin trading on dYdX is link your Ethereum wallet. The Trust Wallet in-app DApp Browser is the easiest method to achieve this on mobile. You may also use WalletConnect to access the web-supported edition of dYdX.

Traders can also use margin trading to short assets they anticipate will fail. Short selling an asset entails borrowing it to sell it and then purchasing it back later (ideally at a lower price) to benefit whenever the asset's value declines.

If Ethereum's bearish anticipates the cryptocurrency's price will fall in the following days, they may utilize dYdX to short ETH against USDC or DAI (with up to 5x leverage).

How to Make Money with dYdX

If you've ever searched through aggregators of DeFi loan rates, you've probably come across dYdX. Because the platform is decentralized, participants may borrow and lend Ethereum-supported assets.

dYdX makes it easier to earn interest on digital assets than any other DeFi protocol. It's straightforward. You will begin to receive interest immediately after you deposit cash into the platform on the 'Balances' tab.

That's all there is to it! And every minute that your dollars are stored in the platform, you will receive interest.

Borrowing (mainly margin traders) on the platform pays interest on stored assets. Because dYdX allows only collateral-based borrowing, if borrowers fail to repay or their collateral value falls below a set limit, their collateral is instantly liquidated to safeguard lenders.

dYdX has reproduced the famous cryptocurrency derivatives offering from prominent centralized exchanges like Bitfinex and BitMEX in a safe, transparent, and trustless manner, rendering it better than its centralized counterparts in many respects.

Uniswap

Uniswap is an Ethereum platform for trading ERC20 tokens. Uniswap, unlike other exchanges, is meant to be a public good—a means for the public to trade tokens without paying platform fees or dealing with intermediaries. Uniswap also utilizes basic arithmetic and pools of tokens and ETH to accomplish the same function as conventional exchanges that connect buyers and sellers to establish prices and conduct transactions.

How It Work

In comparison to centralized exchanges, the Uniswap platform takes a unique approach. Its open-source program is based on the Ethereum network, and it enables participants to trade ERC-20 tokens without the involvement of a third party. This can assist in decreasing expenses while also resolving certain tricky censorship concerns.

Without the use of an order book, liquidity providers play a critical role in making things happen. Anybody can participate in these liquidity pools by contributing the equal value of two tokens, like ETH and stablecoins like USDT or DAI. They'll earn liquidity pool tokens in return, which they may spend on other decentralized applications. This equally assures that they will be able to receive their donation at any moment.

When participants conduct a swap using a trading pair, they are charged transaction fees, with a part of this going to the liquidity provider according to the pool tokens they possess.

Uniswap pools are based on the equation "x * y = k.*."

Here, x might be ETH, y could be USDT, and k could be the result of multiplying x and y to find the pool's total liquidity. Constant product market makers — smart contracts that store liquidity pools — assume that k must always be fixed.

Assume that somebody buys ETH in exchange for USDT via a liquidity pool. As a result of this trade, there will be less ETH in the liquidity pool and more USDT.

Somebody who buys ETH from a liquidity pool in return for USDT will adjust the balance between the assets in this transaction pair, raising the price of ETH while lowering the price of USDT. In bigger liquidity pools, this slippage is typically less evident.

Uniswap's major benefits are its ease of use compared to other DEXs and the fact that traders are no longer accountable for supplying liquidity.

How Uniswap Tokens are Created

The donor gets a "pool token," which is also an ERC20 token, anytime fresh ETH/ERC20 tokens are given to a Uniswap liquidity pool.

When money is placed into the pool, pool tokens are produced, and like an ERC20 token, they may be freely traded, transferred, and utilized in other Dapps. The pool tokens are burnt or destroyed when money is recovered. Every pool token reflects a participant's portion of the pool's overall assets as well as their portion of the pool's 0.3 percent transaction fee.

How to Make Your First Uniswap Trade

You may buy ether (ETH) and any of the hundreds of ERC20 tokens offered by the platform using Uniswap.

To do so, you'll need enough ETH in your account to cover any transaction costs and something to exchange for the ERC20 token you desire. This may be ETH or another ERC20 token. If you want to trade USD Coin (USDC) for UNI, for instance, you'll require USDC in the wallet as well as enough ether to pay the trading fee.

We'll go through how to buy some UNI tokens with ETH and perform your first transaction on Uniswap.

- Step 1: Go to the Uniswap exchange site.

Select the 'Connect to a wallet' option in the top right corner, then log in with the wallet you want to transact with. A MetaMask, WalletConnect, Coinbase Wallet, Fortmatic, or Portis Wallet may be used.

We'll use a MetaMask wallet to log in for this explanation.

- Step 2: The transaction page will display once you've logged in.

Choose the token you like to swap for the other token you want in the top area. You may go with ETH. Look for the token you want to buy in the bottom area, or choose it from the drop-down option, in this instance UNI.

- Step 3: You're finally set to place your order.

You may enter a number in the top area to indicate what you like to spend or a number in the bottom area to indicate what you want to purchase.

Let's assume we acquire 0.1 ETH worth of UNI tokens in our illustration.

- Step 4: At the bottom of the order menu, you'll find an estimate of what you'll get.

Hit the 'Swap' option when you're satisfied with the results. The wallet click would then ask you to authorize the transaction and, if necessary, change the fees to a level that is most convenient for you.

Approve the transaction once you're set, and it'll be executed. The tokens will display in the ERC20 wallet after this is completed.

- Next Steps

There are many choices for advanced users after you've made your first transaction on Uniswap. Because Uniswap is an open smart contract platform, a variety of front-end user interfaces have already been developed. InstaDApp, for instance, enables you to deposit money into Uniswap pools without utilizing the regular Uniswap user interface.

Participants can add funds to Uniswap pools using only ETH rather than ETH and another token utilizing interfaces like Zapper.fi. Easy, one-click options for acquiring pool tokens in conjunction with bZx token schemes are available through the interface.

Balancer

Balancer is a decentralized exchange that uses an automated market maker (AMM) and liquidity pools. It is based on the Ethereum network. It allows participants to trade cryptos and receive interest on their unused crypto holdings.

Balancer participants can trade ERC-20 assets without depending on a centralized body by using liquidity pools. They may also offer liquidity in exchange for a portion of the trading fees. Balancer distinguishes itself from the competition by allowing customers to build their private liquidity pools or pools with more than two cryptos. Balancer provides several incentives to enhance liquidity on the Balancer Protocol in the long run.

Balancer token (BAL)

Balancer (BAL) is the Balancer protocol's native governance token. BAL token owners can vote on the protocol's growth. Although the network does not currently have a defined governance structure, token owners can vote on layer two solutions, the installation of Balancer on blockchains other than Ethereum, and fee adjustments at the protocol level.

How it Works

The Balancer exchange is an automated market maker (AMM), which implies transactions are settled without traditional order books. Balancer pools that are liquidity pools made up of 2 to 8 cryptos are used instead. These offer the liquidity that investors want. The AMM utilizes the percentage of tokens inside a liquidity pool to establish exchange rates. Smart contracts are used to accomplish exchanges.

Whenever a new Balancer pool is formed, the developer determines the token ratio in the pool. For example, if a pool contains Tether, Ether, and wBTC, the ratios might be adjusted to 25 percent, 25 percent, and 50 percent, respectively. When compared to Tether and Ether, there would be twice the amount of wBTC.

When a trader utilizes a liquidity pool to execute a transaction, the liquidity pool is rebalanced. Any imbalances in the pool might provide traders with arbitrage possibilities. The arbitrage trader makes money by taking advantage of the pool's imbalance compared to the actual-world market exchange rate. This arbitrage mechanism constantly rebalances the liquidity pool. However, fees are still collected from liquidity providers and individuals who formed the liquidity pool.

Balancer Pools

To accommodate varied risk appetites, Balancer provides two forms of pools: public pools and private pools. By contributing digital assets to public pools, anybody may offer liquidity to Balancer. These pools' settings are established before they go live and cannot be altered, even by the pool's creators. Stockholders that want to earn fees from their holdings might consider joining a public pool.

Private liquidity pools are those in which only the creator can add or remove assets. The developer also adjusted other pool criteria, such as allowed assets, fees, and weightings.

How to trade on Balancer

A participant will need access to cryptos through a Web 3.0 digital wallet like MetaMask before transacting on the Balancer platform. MetaMask is a browser plugin that serves as a link between the digital asset holdings and decentralized apps like Balancer.

To trade on Balancer, follow these steps:

- Balancer. Log onto the Balancer platform.
- Link your wallet. Link your Web 3.0 digital wallet, such as MetaMask
- Decide on a crypto. Select the crypto you want to sell or purchase from the dropdown set of possible tokens. You may accomplish this by entering the token's name, address, or symbol.
- Fill in the amount. Once you've made your choice, you'll need to input the quantity of crypto you want to trade. After filtering across relevant pools carrying the tokens required, Balancer will offer you the best potential price as you start to input the "token to sell" amount.
- Confirm. On your MetaMask or Web 3.0 digital wallet, click "Swap" and complete the transaction. The crypto sold will be withdrawn from the digital wallet when the transaction is completed on the Ethereum network, and the crypto acquired will be added.

Bancor

The Bancor platform is a collection of smart contracts on Ethereum's blockchain network. Smart contracts on the Bancor platform are used to offer liquidity for traders looking to swap their tokens. On Bancor, anybody may deposit ERC-20 tokens into a liquidity pool and receive interest on their investment. This interest is produced by trading fees on Bancor's platform.

Order books are used by traditional cryptocurrency exchanges to link purchase and sell orders among participants. This functions well for more liquid tokens with a big market value, making it difficult to trade less liquid cryptos.

All supported ERC-20 coins have consistent liquidity thanks to Bancor liquidity pools. Because Bancor is an Ethereum-supported smart contract technology, you can only trade Ethereum-supported tokens on the platform. To trade bitcoin on Bancor, you must first buy Wrapped Bitcoin (WBTC), a form of bitcoin stored on the Ethereum network.

How to Buy Bancor Network Tokens (BNT)

BNT serves two primary purposes: providing liquidity and governing the protocol. BNT is now trading at around $3.52, with a market capitalization of $446 million.

Bancor is a decentralized network that no one party owns. Rather, Bancor is utilized as a governance token for voting on Bancor's network improvement ideas. Participants with BNT have voting rights according to their BNT holdings.

In liquidity pools, the BNT is linked with other ERC-20 tokens. According to Keynes' notion of a global currency, BNT serves as an intermediate currency. For instance, if you wish to swap Ethereum for Wrapped Bitcoin, the smart contract will sell your Ethereum tokens for BNT and then get Wrapped Bitcoin using the BNT you just bought.

- **Create an Online Account**

You don't need to get BNT through Bancor's exchange. Bancor is supported by certain centralized exchanges, including Coinbase, Binance, and Bittrex. Centralized exchanges are more convenient to use, particularly if you currently have a cryptocurrency brokerage account. Most investors who do not wish to use Bancor's decentralized exchange prefer to acquire BNT via Coinbase.

You must provide personal details while opening an account with a bitcoin brokerage. Exchanges usually require your address, Social Security number, mobile number, and email. You only have to link an Ethereum wallet when you use a DEX like Bancor or Uniswap.

- **Buy a wallet (optional)**

It's a smart idea to keep Bancor in a crypto wallet regardless of the exchange you acquire it on. You may store the BNT tokens on the exchange where you bought them, but this is less safe than keeping them in a crypto wallet. Most traders choose digital wallets such as Coinbase Wallet and Metamask, while hardware wallets provide the most protection.

If you plan to put a large sum of money in cryptocurrencies, you must have a hardware wallet. Hardware wallets are physical devices that hold crypto offline, preventing hackers from gaining control over your assets.

- **Make Your Purchase**

After you've decided on a software or hardware wallet to hold your cryptocurrency, you'll have to buy Bancor Tokens. You have the option of using a centralized exchange such as Coinbase or Binance or a DEX such as Bancor or Uniswap.

The convenience of centralized exchanges is one of its main advantages. All you have to do is open a brokerage account and put in a purchase order after that. You'll require an Ethereum wallet extension for the Google browser to use a DEX. MetaMask is a prominent wallet extension.

You may put your order as a limit or market order, like when buying a stock on the stock exchange market. You select the rate you want to acquire BNT when you make a limit order for BNT, and if a counterpart sell order is made at that rate, your order will be completed. Instead, you may use a market buy order to purchase BNT at its current market price.

- **Trade or Sell Your BNT**

Once the buy order is fulfilled, the BNT will be transferred to your account. You may move the BNT to any Ethereum wallet via your wallet's address after your tokens have been transferred to your account. If you intend to trade your BNT constantly, you may wish to retain it on the exchange where you bought it. On Ethereum, you must pay a trading fee known as 'gas' each time you make a trade.

If you're going to invest in BNT for a long time or with a large sum of money, you need to use a hardware wallet to ensure your fund is safe. There have been numerous crypto hacks in the past, and storing your assets on an exchange puts them at risk.

3

YIELD FARMING

Meaning of Yield Farming

Yield farming (YF) in DeFi is now one of the most popular trends right now, providing investors with an even better chance to boost their profits.

So, what exactly is DeFi yield farming? Yield farming is a practice in the DeFi crypto industry. It is the word used to describe the process of achieving the best yield and a way of earning more crypto with your existing crypto. It's also an opportunity to earn additional payouts from the protocol's governance token.

Traditional traders view crypto yield farming as akin to bonds and dividends. The yield on DeFi tokens varies based on how different initiatives distribute them. If the price of each asset rises, the yield given on your crypto offers consumers more tokens, which cost more money, similar to dividend distributions. This incentive scheme has piqued the curiosity of millions of current merchants.

Some analysts have compared this phrase to bank loans. When a bank borrows money from a customer, it must repay the amount with interest. Banks are crypto owners. To offer liquidity in DeFi protocols, yield farming uses "idle cryptos" that would otherwise be squandered in an exchange or hot wallet.

Principles of Yield Farming

So how do YF works? First and foremost, consider a market for DAI and USDC. Every time you see one of these tokens, it's worth $1. To create a USDC/DAI pool, start by contributing equal amounts of both tokens. In a pool with only two DAI and two USDC, a single DAI would cost one USDC.

The pool would then have one USDC and three DAI. The pool would be completely unbalanced. By investing in a single USDC and receiving 1.5 DAI, a user might make 50 cents. There is a 50% arbitrage profit and also the issue of limited liquidity. A transaction of one DAI and one USDC will have no impact on the relative cost if 500,000 DAI and USDC of the same quantity. It facilitates the utilization of liquidity.

Comprehending how YF works also necessitates a basic knowledge of smart contracts, as they play an important role. Smart contracts, which are essentially little software programs, operate as a link between your money and the money of other participants.

Earning profit is the goal of any form of lending, and cryptocurrency lending is no different. For their services, a lender receives fees in the form of token. YF is one of the most common ways to profit from crypto ownership. DeFi YF follows in the footsteps of auto-market makers (AAM). One of the several decentralized exchange protocols, which includes liquidity pools and providers.

Now that we've talked about the foundation let's look at how the whole thing plays out.

- Stage 1:

At this stage, smart contracts serve as liquidity pools. Money is deposited there by providers. The smart contracts lock stablecoins, a completely new type of cryptocurrency that promises to provide stable pricing and is supported by a reserve asset. The contracts unlock stablecoins under specific limitations and YF platforms.

- Stage 2:

Participants may sell, lend, or borrow YF coins with this app. The users are required to pay a fee. Market makers are paid a return on their investment depends on the quantity of money they put in.

- Stage 3.

At this point, market makers are rewarded for their readiness to put money in the pool. The incentives that participants receive are determined by protocol, and the quantity of money contributed.

- Stage 4:

Providers reinvest and transfer their incentives to boost their profits even further. To put it another way, they continue to store currencies in liquidity pools. Liquidity providers use this method to expand their investment portfolios and raise money. Selecting good strategies allows you to maximize your profits.

Another element that determines how much a user may make is the pool's activity. Stablecoins tied to the US dollar, such as DAI, USDT, and BUSD, are preferred. You could be wondering exactly how much you can earn in return now that you understand how YF works. You may find instructions on how to estimate your profits in DeFi YF below.

Estimating Returns in Yield Farming

Learn how YF returns are estimated using certain indicators and formulas. Using a YF calculator, you can calculate this parameter every year. It will show you the potential profits over a given length of time. APY and APR are usually sufficient, but some participants use a third indicator, total value locked. So, here's how to figure out how to estimate LP returns.

- Total Value Locked (TVL)

TVL is a metric that indicates how much crypto has been locked up in DeFi lending and other money markets. This metric may be used to assess the state of the YF environment. The "golden rule" appears to be as follows: the higher the value locked, the higher the yield farming that is likely to occur. To evaluate this parameter, use USD, ETH, or BTC. The level of interest you pay each year is expressed as an annual percentage rate (APR). APR overlooks the influence of compounding.

- The Annual Percentage Yield (APY)

APY is the rate of return you get on your money. However, compounding interest is vital in this situation.

Compounding is a term that many people are unfamiliar with. Simply put, it refers to the practice of immediately reinvesting profits to generate even more profits. It is possible to use APY and APR synonymously.

Even yet, forecasting the outcome of random table games such as keno or bingo is nearly as challenging as evaluating ROI in this industry. Those computations would never be correct. This is because yield farming is a highly competitive and fast-paced industry. As a result, the rewards might change at an incredible rate. Most people would use a certain strategy if it proved to be beneficial over time. As a result, high returns may be avoided.

It's important to remember that YF isn't free of capital losses. Virtual trade protocols between a pair of unknown individuals are used in YF transactions, and there is no central independent authority. Financial data is in danger of being lost if a system problem occurs. The records would be safe if blockchain prohibits all types of system delegation.

How to Obtain ROI

In the yield farming process, ROI may be grouped into three different types. They include:

Token prizes. These are similar to incentives to offer liquidity. The tokens are given over a certain length of time, which might range from weeks to years. Participants may trade such cryptocurrencies on decentralized exchanges as well as other exchanges like Binance and Coinbase. The use of these tokens is intended to control a system.

Transaction fee revenue. Consider that the commission offered by the user during pool formation varies between 0.003 percent and 15%. That is the rate of one particular pool. Others request a 0.02 percent fee. All fees are paid to market makers. Owners of governance coins will most likely receive a portion of the earnings in addition.

Raise of capital. An increase in money makes it easier to calculate the revenues of any YF opportunity. When REN, BTC, CRV, and SNX assets are included in a YF approach, they become extremely volatile and no longer need correlation. To avoid volatility, it is important to use suitable techniques that are compatible with stablecoins.

When working with YF, this is how you receive a return on investment. Let's go on to the most effective yield farming protocols to work with.

Yield Farming Platforms

You may now wonder how you can profit from this idea. Bear in mind that there are a variety of yield farming strategies available, and new ones emerge on a routine basis. It's impossible to use all of them at the same time. It might also be difficult to recall them all by heart.

To put things right, research each platform you want to use to see which strategies it advises. Furthermore, have a broad understanding of how decentralized liquidity protocols function — this should suffice for your first time.

Any expert will warn you that you should avoid mindlessly putting money on the first website you come across. Also, be aware of the risk management guidelines. To make things easier for you, we've compiled a list of popular yield farming protocols. You can find them listed below.

Compound Finance

Coins can be borrowed or landed here. To begin, you must create an Ethereum wallet. Only with this set will you be able to begin receiving rewards that will compound over time. Compound Finance automatically adjusts rates based on supply and demand. This protocol is one of the most used.

Compound's current incentive system for farmers, via the issue of its native governance token COMP, is among the most enticing features. In reality, everyone on the Compound network who loans or borrows can farm a particular amount of the COMP coin.

Compound has given out 976,102 COMP Tokens to farmers and borrowers thus far. At the moment, 2,312 COMP tokens are given daily among the Compound participant base, resulting in more than $920,000 in extra incentives every day for $400 per COMP token. Although the COMP farming incentives are spread among the platform's 294,000 suppliers/farmers and 8600 borrowers, the motivation for participants to farm on Compound is quite great, albeit the comparatively modest APYs.

Compound also includes its native tokens, known as cTokens, which are intended to compensate farmers for providing liquidity to the system. For example, when farmers deposit and lock 5ETH on Compound, the protocol creates 5cETH tokens, which collect interest and may be re-deployed on other DeFi protocols. Farmers may then exchange their cETH for ETH, as well as their staking rewards, at any time.

Participants will have to do the following to engage in YF on Compound, and also most other YF platforms:

- Buy cryptocurrency for use on the specific YF platform. ETH, BTC, and stablecoins like DAI, USDT, USDC, and BUSD are all popularly recognized cryptocurrency assets (for farming).
- .Download a wallet, Metamask, Trustwallet, or Wallet Connect are examples of decentralized wallets. As soon as you're requested to register, ensure your private keys and seed phrase are safe and protected. You may achieve this by following Guy's step-by-step instructions.
- Send money to your favorite wallet after you've installed it.

To begin farming, go to the wallet's dApp area. Due to the COMP incentive and its simplicity, it is recommended that new farmers begin farming using the Compound platform. In the case of Compound, people who want to farm should:

- Visit Compound Finance
- Select 'App'
- Compound App
- Then connect wallet via the Metamask icon.
- Connect Your Wallet
- Approve the password-protected connection and unlock the wallet.
- Unlock Wallet COMP

Participants can pick from a list of assets to give after the connection has been authorized to begin farming COMP. Participants must first select 'Collateral' and afterward 'Use DAI As Collateral' if they like to furnish the platform with a stablecoin like DAI.

.DAI Collateral should be enabled. Users must enable DAI as collateral and pay a modest ETH trading fee after that. Participants will place their DAI onto the Compound platform and begin farming COMP after the transaction is completed. The Participant's APY, as well as there "Supply Balance" and "Interest Received and Paid, plus COMP," will be shown in the "Dashboard" area. DAI should be deposited in the compound. It's equally worth noting that the more assets a farmer has, the more borrowing ability they have.

Uniswap

Risky currency exchanges are possible on this platform. To create a market, liquidity pool users need to deposit the equivalent of a few tokens. They get their fees from pool trades in this way. Most investors are drawn to Uniswap because of its frictionless quality. When experimenting with various YF strategies, keep this platform in mind.

Uniswap is probably the largest liquidity pool in DeFi, since it is among the industry's most well-known Ethereum-supported AMM protocols. Liquidity Providers (LPs) can earn fees by contributing money to a pool through Uniswap. Liquidity pools on Uniswap are split 50-50 between two assets, as with Automated Market Makers (AMMs).

LPs are critical to Uniswap's ability to function as a decentralized exchange since they offer the liquidity and collateral required for the protocol to conduct transactions decentrally. In reality, every time a person uses a liquidity pool to complete a transaction, the LPs who contributed to the pool get paid a fee for enabling the transaction. The exchange costs a 0.30 percent trading fee for each coin swap. However, instead of going to Uniswap, these payments go to Liquidity Providers to thank you for supplying capital.

Adding Liquidity On Uniswap v.3

Uniswap, unlike other decentralized exchanges, does not have order books and instead relies on liquidity pools to keep its liquidity. This implies that anybody may join Uniswap as a liquidity provider (LP) for a coin pair by placing equal quantities of every coin in return for token pools. For example, if a participant wanted to contribute liquidity to an ETH-DAI pool on Uniswap, they would need to add an equal amount of every coin.

Structure of the Uniswap Pool

1 ETH is equivalent to about 2,270 DAI. For example, if the LP wished to add liquidity to the pool with 3 ETH, the required 50-50 ratio would be 3 ETH – 6,810 DAI.

Participants must do the following to contribute liquidity to a Uniswap pool and begin YF on the platform:

- Go to Uniswap.org
- Click the 'Launch App'
- Select 'Pool'
- To connect with Metamask, click the 'Connect Wallet' button.
- After connecting, participants may either explore prominent liquidity pools by selecting 'Top Pools' or start a new position by selecting 'New Position.'

- LPs can choose their favorite token pair after clicking on "New Position." After that, they must reconsider their desired Fee Tier.
- Examine the Fee Tiers

It's worth noting that Uniswap v.3 offers three distinct Fee Tiers for each token pair: 0.05 percent, 0.3 percent, and 1.0 percent. The 0.05 percent Fee Tier is appropriate for assets like stablecoins that trade at a fixed or strongly linked rate. As a result, this Fee Tier is best suited for liquidity pools like DAI-USDC or USDC-USDT.

The 0.3 percent Fee Tier is ideal for most pairings, especially those that see price swings, such as ETH-DAI. This higher Fee Tier is more likely to reimburse LPs for the higher price risk they assume compared to stablecoin LPs. The 1.0 percent Fee Tier is designed to compensate LPs for taking substantial price risks on their holdings.

- Determine a price range

One of the benefits of the new Uniswap update is that LPs may specify a certain price range to supply liquidity. This implies that if prices go beyond the specified range, the participant's holding will be focused on one of the two assets, and no interest will be paid until prices return to the range.

- Deposit Uniswap Tokens. Place the required token quantities in your account.
- On Metamask, you may 'Add, ''Preview, 'and 'Approve' transactions.

MakerDAO

The DAI stablecoin is supported and maintained by the decentralized credit service. Maker Vaults can be created by anyone who selects this protocol. Stablecoin may be generated as a debit on locked collateral funds like ETH, BAT, WBTC, and USDC. This protocol assists in the selection of the best YF strategies.

Synthetix

It's a blockchain-based distributed asset issuance system that is based on the Ethereum blockchain. This platform stands out from the rest due to a few unique characteristics. Synths may be created and converted by anyone. Peer-to-contract (P2C) trading implies that all transactions are completed quickly and without the need for an order book. Any Synth may be traded for another Synth, and this protocol provides near-infinite liquidity. A group of token holders is responsible for providing collateral and ensuring the service's stability.

Aave

It's a fantastic way to borrow or lend money. Interest rates are automatically adjusted in response to market conditions. This protocol has created native coins that will be used to reward users. When you invest money in aTokens, you get immediate returns and compound interest. Those that make larger deposits receive more aTokens. Aave also offers short-term loans. This, along with a slew of other appealing features, made this system extremely popular.

Yearn Finance

This platform proposes the highest returns on ETH deposits, as well as the finest altcoins and stablecoins. This platform turns yTokens into invested coins. Yearn Finance's smart contract looks for the greatest APR YF DeFi protocols to deliver money there.

Curve Finance

This technology allows for high-value stablecoin trades to be conducted with little slippage. Pools are an essential element of the architecture of this DEXs system. Because of the availability of stablecoins in the yield farming sector, this is the case.

Balancer

What distinguishes this platform from others? Rather than the 50/50 distribution claimed by other platforms, it enables the creation of bespoke Balancer pools. Custom currencies can be assigned to a liquidity pool by providers. The transactions that happen in liquidity pools profit liquidity providers. Balancer has a lot of versatility, and it's a fun approach to look at yield farming strategies.

Yearn.finance

This protocol makes every effort to improve coin lending by locating successful loan businesses. Earnings are rebalanced and raised once liquidity providers make a deposit. This protocol chooses the most successful yield farming strategies for you.

These are the best YF platforms you can rely on right now.

Pros of Yield Farming

As you have seen, there are many solid reasons to consider yield farming as an investment option. Compared to traditional approaches, yield farming is likely to become an ideal market with several chances to uncover high return rates. That is, as bitcoin becomes more widely accepted, the need for crypto-based financial services will increase.

But it's also a complicated strategy. As a result, if you plan to participate actively in the cryptocurrency world, you should research it well. You may do it yourself or use OpenGeeksLab. A recognized DeFi YF development firm.

One of the most significant benefits of YF is that it is one of the most effective options for saving money in the account. Yield farmers may make far more money via YF than they might from regular banks. Holders with idle funds may invest in DeFi protocols to generate additional cryptos, making it an excellent source of passive income.

DeFi yield farming opens up profit opportunities for both users and platform owners. It's easy to see why: this process has numerous benefits. Let's concentrate on the three most important.

- Simple User Interface

There are many YF tools available to help you stay on top of your assets.

They have a relatively short training time. Their user interface is easy and enables you to review project availability and deposit cryptocurrency.

- Easy to Start

This is due to the user-friendly interface once again. You don't need to be tech-pro to get started; special tools will take care of everything. The high interoperability of decentralized finance services also ensures an easy start. Having a cryptocurrency wallet and Ethereum are the two most important prerequisites; in most cases, these are sufficient.

- Profit Potential

People that decide to invest in protocols early, very much like cryptos, can make a lot of money. To put it another way, a strong return on investment is what appeals to most investors.

This list of advantages is not exhaustive. Yield farming has gotten a lot of attention because it's one of the most profitable forms of cryptocurrency investment with huge liquidity. Simplified regulations and a growing number of users are allowing yield farming to flourish.

Cons of Yield Farming

- Short-Term Rewards

It is undeniable that YF is gaining traction as a fast-paced industry, but the market remains unpredictable, posing a danger of uneven profits. Furthermore, lucrative methods are difficult to figure out due to the accessibility of entrance.

- Ethereum's High Gas Fees

Simply said, gas is the cost charged for every trade on the Ethereum blockchain network. Gas prices have been on the rise recently, and this is one of the drawbacks of YF. As a result, users should avoid paying gas prices that are larger than the projected benefit.

- Benefits for Those With More Capital

While DeFi enables anybody to engage in YF, the rewards will be significantly larger for those who start with a large amount of money. This is because the more cryptocurrency you possess, the more you may invest in high-yielding methods, resulting in a larger return.

- Impermanent Loss Risks

This relates to a liquidity provider's (LP) transitory loss due to a trading pair's volatility. Impermanent loss is a key stumbling block for AMM protocols. It happens when the prices of coins inside an AMM deviate too fast in either direction, resulting in coin instability.

Risk Associated with YF

YF, like any other business, has its advantages and disadvantages.

- Liquidity Risk

This risk arises when the collateral value falls below the loan fee, resulting in a sanction on the collateral. Liquidation happens when the value of the collateral drops or the value of the loan rises.

- Price Risk

Assume that a participant has amassed a significant amount of value from the service due to their yield farming strategies (for instance, 210 percent). This individual will lose because the price of the token has dropped in the market. If the collateral value falls below a certain threshold, the borrower will be removed from the platform before they have an opportunity to repay their debt.

- Strategy Risk

Assume that loan pools have become swamped as a result of reduced liquidity. Arbitrage trading is the process of examining a variety of exchange services to profit from price inefficiencies. If volatility falls, arbitrage trading will be no longer profitable.

- Scam Risk

Because creators have power over the currency, there's a risk that they'll flee with it. The risks are much greater if you don't know the creators. Users must always ensure that the pool they choose has been thoroughly researched by a team they trust, but this does not completely remove all risks.

- Gas Fees Risk

During the apex of the DeFi season, gas fees increased by about 100 times. If they continue to rise, yield farming may become unaffordable for most investors. ETH has launched Ethereum 2.0, which includes layer II scalings to address increased gas fees. Reduced gas fees are also available from BNB, NEO, and TRON.

So, while there are two sides to the coin, we think you should seize the opportunity to try yield farming and concentrate on the rewards it may offer. To avoid issues, bear in mind all possible risks sources.

4

DEFI RISK MANAGEMENT

Risk Management Strategies

The crypto industry has been completely changed by decentralized finance. Investors are attracted to this industry because it is both a potential infrastructure project in the cryptocurrency business and a way to benefit from volatile token fluctuations.

Simultaneously, this volatility comes with significant risk. Let's take a look at a couple of DeFi market strategies that might help mitigate these risks. In 2021, general funds invested in DeFi protocols increased from $20.1 billion in the first quarter to $120.3 billion in June. Furthermore, the mean daily trading volume on DEX, a key component of DeFi, increased by 100%. It surpassed $2 billion in Q1 2021 but stayed below $1 billion in Q4 2020.

DeFi provides traders greater power over their money. Therefore investors have been drawn to it from the start. While DeFi began with loans and credit, it has expanded to include at least five fully functional, linked segments: blockchain network, decentralized exchanges (DEX), loans and credit, decentralized derivative networks, and insurance.

Despite its overall stability, the decentralized finance sector is nevertheless a risky business. The high volatility of numerous project coins undoubtedly draws new investors. Nevertheless, a high return also entails a high level of risk. In decentralized finance, there are some successful risk management strategies.

Diversified Portfolio

Using this strategy, traders stockpile the most attractive, inexpensive, and non-overlapping decentralized finance projects in their portfolios. This reduces the chance of a single item losing a substantial amount of value in the portfolio. The market cap to total value locked (TVL) ratio may be used to discover potential coins. Coins with the least correlation of these values might be deemed cheap, implying that there is reason to think that their valuation will "catch up" with others and rise. Investors may thus maintain their portfolios for the medium and long term by purchasing these coins.

Pick projects on multiple blockchain networks, like Ethereum or Binance Smart Chain, to further spread the risks. Another approach is to use negatively correlated coins, which means that if one coin is very unstable in day trading, you could balance it with a more stable token. The Uniswap (UNI) and Zilliqa (ZIL) tokens are both relatively stable assets.

It's even smarter to diversify the portfolio by including an insurance project because the need for capital protection in the DeFi ecosystem is only starting to skyrocket. Cover, Nexos Mutual, Etherisc, and Opyn are some projects that offer decentralized finance insurance. If an unfavorable event occurs unexpectedly, holding such tokens in your financial portfolio will greatly enable you to mitigate the impact.

Staking

With stake-able coins, you may also reduce your chances of losing money. Some decentralized finance projects' coins allow users to earn just by keeping them. Participants of the worldwide crypto exchange CEX.IO, for instance, may get up to 16% yearly interest on their coins. And whenever the value of the currency grows, so does the reward from staking. For instance, when a trader buys $100 in ZIL at $0.20 per token and earns a 16 percent staking interest, they will get around 580 ZIL at the end of the year. If the price has grown from $0.20 to $0.40 during that period, the trader will receive $232 when they cash out. The profit increases to 132 percent rather than 20 percent when the token's price increases.

You may buy stablecoins that allow staking, like Dai, to safeguard your portfolio from significant volatility. As a result, the earning from your tokens may both boost your relative profit from investing in the decentralized finance sector and protect you from any losses if the price collapses. Investors can also restrict their losses to the level of their staking profit possibility. Investors can easily terminate their positions immediately after the losses on their holdings approach the gain they would earn from staking.

Hedging

When investors purchase an asset on an exchange in the standard hedging structure, they instantly establish an opposite position in the related derivative. Futures, options, and contracts for difference are examples of these derivatives. Therefore, when an investor buys UNI on an exchange, they will sell a contract for difference for an equal amount to hedge the risk of the asset decreasing in value. Finally, if the value of UNI rises, the losses on the contract for the difference will be compensated by the rise in the prices of the crypto. If the value of UNI falls instead, the difference will be offset by the CFD's opposite position.

Certain platforms provide derivative trading, and CEX.IO Broker is one of them. You may use it to benefit from value changes in crypto without the need to acquire them in person, as well as to hedge assets you own. A contract for difference enables investors to escape the technical risks, volatility, and complexity present in DeFi by participating indirectly. Automatic safety measures, like stop-loss and take-profit, are used by the platform to manage risks.

CEX.IO has built a complete trading and asset management environment. It enables participants to profit from decentralized finance market movements to the greatest extent possible and, if required, employ techniques to protect their funds successfully.

Right away, participants are presented with a complete solution. There's portfolio diversification: there are a lot of decentralized finance currencies to invest in and trade. Staking, for instance, provides a greater yield than the network rewards: for example, ZIL staking pays 16 percent interest, but the network pays just 14.2 percent. Lastly, at CEX.IO Broker, decentralized finance coins may be traded as derivatives.

CEX.IO Broker's versatility allows you to establish up to ten accounts as part of a single user account and test alternative strategies separately. A demo account is available for people who have never dealt with the cryptocurrency market previously. Investors can go to a genuine account and begin trading the currency pairings they choose to be sure of their abilities. CEX.IO Broker is a margin trading platform that allows you to begin trading with less money than you would with spot trading. This allows you to leverage your trading capital to raise your trading capital.

Types of Risks Decentralized Finance

Financial, procedural, and technical risk are the three most frequent forms of risk in decentralized finance. Financial risk is concerned with the possible benefits of investment possibilities as well as their management. Financial risk is sometimes ascribed to an organization or a user's risk profile. Financial risks are also influenced by a user's goals for managing an effective investment portfolio.

The technical risk of decentralized finance platforms or services is specifically related to hardware and software problems. Participants and the strategies they employ to use decentralized finance products or services provide procedural risks that could undermine security. Procedural risks are nearly identical to technical risks, except the end-user relationship.

Technical Risks in Decentralized Finance

Protocol, hardware, and software problems are the most common sources of technical risks in decentralized finance. Technical risks are a major concern since they can jeopardize the system's overall operation. Race situations, API, use cases and exception handling, I/O handling, and memory security are all examples of technical risks. In most cases, a race condition renders the series accountable for an event's result inaccessible.

Memory interruptions, access failures, uninitialized variables, and memory risks are all issues to consider regarding memory security. User experience could be harmed by a lack of appropriate testing for use cases and exception handling. APIs' features are also hampered by a lack of adequate testing and assessment. Lack of adequate testing exposes processes to technical risks as a result of I/O handling among systems.

Smart Contracts, Hardware, and Software Risks in Decentralized Finance

The technical risks in decentralized finance are also influenced by smart contracts, software, and hardware. Due to the significance of smart contracts in allowing automation, smart contract risks in decentralized finance are important. Smart contracts also have some flaws that pose a technical risk to decentralized finance.

Dependence on a timestamp, front-running, insufficient gas griefing, integer underflow and overflow, and forceful ether transfer to a contract are risks associated with smart contracts. Attackers could use transactions mempool to grab an unincluded block and modify it to their liking, posing a front-running threat.

When the code cannot restrict the size of the unit variable to 2256, there is a risk of integer overflow and underflow. The value is immediately adjusted to zero if this occurs. When miners try to change the timestamp of a block, they run the risk of becoming dependent on it. The smart contract is prone to self-destruction when ether is forcibly sent to it. Gas griefing hazards associated with smart contracts result from starting transactions without paying attention to the transaction sub-call.

Hardware threats are equally significant technical concerns in decentralized finance, particularly because hardware serves as the backbone of the architecture of decentralized systems. Sensitivity, power problems, and incompatibility are all major hardware risks associated with decentralized finance systems.

Voltage variations represent a danger to service life and effectiveness, while power problems may cause instability in the service or app. Degradation, humidity, dust, and other related problems can cause hardware to become sensitive. Incompatibility risks hardware refers to hardware drivers that might slow down the system and result in other problems.

In decentralized finance, one of the most important technical risks is software risk. DDoS (Distributed Denial of Service) assaults, injection, uncontrolled format strings, and overflow are potential threats to decentralized finance software. DDoS is a legitimate approach for interfering with an application's or service's normal operation.

Injection risks indicate the possibility of harmful code being introduced into decentralized finance software, with SQL injection for online applications being one of the most common injection risks. Uncontrolled format strings are dependent on forms, which can run malicious malware in a web project. In decentralized finance software, overflow risks manifest themselves in skipping some software functions or their implementation in an unfavorable manner.

Financial Risks in Decentralized Finance

Financial risks are the second most crucial type of risk in decentralized finance. The financial risks associated with DeFi give information about making better use of DeFi protocols and services. Developers, for instance, must concentrate on doing the right thing and lowering financial risks for users by providing accurate advice and implementing changes in their DeFi app.

Financial risk refers to the possibility of losing money, and each participant is responsible for comprehending financial risk based on their benefit and risk propensity. By regulating fund management based on company activities, a company, on the other hand, will concentrate on financial risk.

Additionally, financial risks in government authorities are dependent on the administration and allocation of cash across multiple systems and solutions. Decentralized finance's universal design makes it a good fit for all of the above while assuring appropriate increases in value. As a result, it is appropriate to employ risk management methods like technical analysis and fundamental analysis in financial planning and investors and innovators in the DeFi ecosystem.

Fundamental analysis uses numerous measures and ratios to evaluate the value proposition of various assets. Consequently, fundamental analysis exposes both the commercial value and the financial performance of a company. The technical analysis complements fundamental analysis by employing quantitative metrics, graphs, and trends to comprehend better the risks associated with a particular investment.

Procedural Risks in Decentralized Finance

Procedural risks are highlighted in the final risk entry. Procedural risks, on the other hand, are largely concerned with the various security problems connected with DeFi platforms and services by participants. Phishing issues, whereby a malicious agent simulates a website or service to trick unwary participants into giving important details, are the most prevalent security risks in decentralized finance.

Phishing attacks may equally be carried out via emails, whereby participants are given an email that looks just like that of a service provider. The participant is routed to a malicious website immediately after they open the email. The phishing email, contrarily, might install keyloggers in the user's system by running malicious code in the browser.

The hacker might then utilize sensitive data to move funds or carry out criminal operations without the participant's awareness. Hackers acting as personnel of a decentralized finance service are common in phishing attacks in the crypto ecosystem.

Other significant procedural hacks need to be accommodated in the comprehension of procedural risks in decentralized finance. Baiting, pretexting, SIM-swapping, spearfishing, quid pro quo, and tailgating are all potential threats. Pretexting is when a hacker poses as a DeFi service agent and persuades participants to disclose important information. Bait and switch techniques for infecting a web page pose a danger of baiting.

Spear phishing poses a risk to the general business since it targets specific people to access the system. Spear phishing is getting system access from anybody to manage the system's essential functions and data. The risks of Quid Pro Quo are similar to those of baiting, except that hackers provide substantial incentives to motivate users to act in line with their plans.

SIM-swapping is a common procedural risk associated with decentralized finance, owing to participants' sensitive information to create new SIM cards from affected mobile service providers. Hackers may use the fake SIM to carry out criminal actions under the user's identity. Tailgating is one of the most common risks in decentralized finance when gaining access to actual-world places by deceiving someone in a superior position.

Practical Tools in Decentralized Finance Risk

- Use data in Ethereum blockchain with Etherscan.

Etherscan, an Ethereum Blockchain network Explorer website, is a handy tool for examining transactions and the condition of the Ethereum Blockchain network. It may be used to look up and verify any trade or transaction on the blockchain network. The majority of blockchain networks have their explorer.

It is advised that you use Etherscan to monitor the wallet transactions and input your Ethereum wallet address. Examine the recipient addresses of any recent transactions you've done, and if there are any odd situations, like any bizarre behavior, address them as soon as feasible.

- Use Coingecko to get the most recent information on cryptocurrencies as well as the most recent price changes

This is a very handy tool for getting full and reference information on different currencies, as well as price fluctuations and market capacity. You may also utilize Coingecko to get the most up-to-date information on liquidity and Farming Pools, as well as the current APY, the number of audits linked with that platform, and the total value locked.

You can get the most detailed information about liquidity pools in various blockchain networks by utilizing the LiquidityFolio platform. You may also verify your presence in multiple pools by inputting the wallet address into this platform.

- The Zapper platform aims to make fund management easier in decentralized finance.

A Defi fund management program aims to deliver services on decentralized finance platforms so that all assets may be viewed in the dashboard. Furthermore, the platform aims to simplify it by offering the most significant liquidity, Farm pools, and relevant filters.

- Use the Zerion platform to conduct all key activities in Defi via a good user interface.

It is a useful tool for controlling all critical processes in the Defi ecosystem. Using this platform, you may track your portfolio in a chart format that is very useful and thorough in presenting portfolio assets. It even keeps track of NFT assets. The fact that this platform has a nice user interface is one of its most significant features.

There are additional parts for Exchange, Borrow, and a search bar for key liquidity pools and YF sites on the network. Another benefit of these platforms is their partnerships with DEXs and the simplicity they may conduct transactions.

For instance, during the initial 1inch Airdrop, the 1Inch platform that distributed 600 tokens to its participants had some initial delays owing to increased traffic from participants claiming coins and requesting transfers to their wallets. Participants that checked the 1Inch Twitter account and learned about the system's connection with Zerion, on the other hand, were able to move Airdrop coins to their Zerion wallet quickly.

- The YF Tools platform offers significant Defi-related tools and resources.

This tool includes features like estimating the Impermanent Loss, the size of Collateral, and more. One of the benefits of this platform is that it allows you to learn about the most significant and current issues in decentralized finance, which can be found in the Resource section.

You may use the Impermanent Loss estimator equally and obtain the most recent Ethereum gas Fees under the Tools section. This platform includes a liquidity pool search bar. However, it differs in that it has extremely great and unique filters. Filters like APY rates, smart contract risk, and Impermanent Loss, as well as numerous platforms which provide decentralized finance, collateral type, and coins in Pools, and benefits for participation in different Pools, are all extremely helpful.

- Trade through the @GaspriceTrackerBot bot on Telegram, and set notification for the right time.

This bot is designed to give recent information about Ethereum Blockchain network transaction costs. This bot can also send out notifications when fees are low or warnings when the gas fee charges are high.

- Use the Gasnow site to get the optimal gas fee for you based on the time priority that is most convenient for you before conducting any transactions on the Ethereum blockchain network.

This site, which is backed by Sparkpool, one of the largest Ethereum network mining pools, provides the most exact amount of gas fees in four distinct periods. Another benefit of this site is that it provides a helpful chart of the amount of gas fee vs the number of completed transactions that can forecast the traffic on the Ethereum blockchain.

- Determine the risk profile of smart contracts and helper tools by visiting Defiyield.info

It is with no question among the most crucial decentralized finance management systems. This tool may look for liquidity pools and Vaults from other platforms to invest in. One of the essential aspects of Defiyield is that it displays the audits that have been completed and the risk level of smart contracts in the decentralized finance landscape.

It equally contains the Impermanent Loss estimator, which has the most precise and user interface. Stopping Ethereum Network Transactions and a Gas Cost Monitor are two other significant features of this platform.

- Use lending management features with the Defisaver platform.

It's among the most crucial instruments in the loan and borrowing landscape. We could effectively safeguard our Collateral against loans acquired from various sites due to this technology, giving us peace of mind.

We may also return our loan by generally making just a single transaction, which saves us money on trading costs. Another helpful function of this platform is the ability to transform collaterals to one another or move your holdings from one network to another in a single transaction. In addition, when a participant launches Smart wallet on the platform, a Maker CDP is immediately generated for them.

You may equally apply the Loan Shifter function after developing a smart wallet. In brief, DefiSaver lowers trading costs in the lending and borrowing landscape while also making fund and collateral management super simple.

- Benefit from the new decentralized finance services in lending by using the Instadapp platform.

The platform has also shown to be highly beneficial in the lending industry and strategy execution. Instadapp is recognized for its unique approach to participant service in this domain.

The use of Uniswap LP coins in leverage and collateral strategies, for instance, is highly intriguing and creative in the most recent upgrade of its services. Professional decentralized finance participants have been able to boost their revenue via the fees of AMM platforms thanks to this innovative approach.

- Furucombo allows you to do several transactions in one.

The platform is intended for participants already acquainted with Defi and want to make complicated hybrid transactions easier. Furucombo allows you to perform a set of complicated decentralized finance protocols strategies in one transaction. You may equally gain ideas from other example combinations in the Explore area and utilize them to optimize your trades.

- You may browse and get rapid and classified access to decentralized applications by utilizing dappradar.

With the proliferation of decentralized apps on various blockchain networks, classified access and exploration using realistic filters are becoming increasingly important.

This platform serves as a conduit between blockchain network developers and participants. Statistical reporting, fund management, and recent information on NFT markets are among the other advantages of this platform.

5

DEFI PORTFOLIO MANAGEMENT

Overview

We're nearing the conclusion of our detailed examination of the many macro aspects of decentralized finance (DeFi). This time, we'll concentrate on portfolio management and the networks that enable interaction with them.

The topic of the fund and the systems that enable them to be managed is equally fascinating. Under this section has the different coins tied to other assets, like all those connected to Bitcoin (BTC), systems, and platforms that enable you to manage them. Including other digital assets, like Idle Finance.

In fact, over the period, protocols and smart contracts have emerged that have created the opportunity not only to enhance and digitize certain aspects of trading but equally to manage digital funds of different assets. This enables you to trade assets based on market trends and achieve the best possible result based on user-defined metrics.

Consider some price trend graphs and their median values, which form a more or less regular trend, and then deciding if to sell or appear the relative asset and make a gain, or restabilize with regard to some other asset, and then selling one to acquire another or converting into a stablecoin to hold for the right time to get back to operate.

Even though the numerous interfaces are basic, it is tough to comprehend how to use and operate these tools, particularly for novices or those new to this industry.

These tools, as amazing as they may appear, usually conceal the issue of being appropriately managed because the market situation could change suddenly, resulting in both tremendous profits and woeful losses, as well as the loss of all funds. Therefore, they should be used to understand cause and control, which does not allow you to feel completely relaxed.

We monitor our assets ourselves without the assistance of third parties that can take our place and mentor us through the various options. This can be advantageous because we are not subject to the will of a third party, but it also means that we are solely responsible for our actions and have no recourse if we make mistakes.

Need for DeFi Portfolio Management

With novel financial solutions, the number of platforms based on decentralized finance protocols has risen. The many tools for managing, controlling, and hedging risk across a variety of decentralized finance products are certainly impressive. They've proved to be effective in a variety of projects, including lending, derivatives, and DEXs. Decentralized portfolio management solutions can change the traditional consequences of fund management by providing trustless community, composability or interoperability, and transparency.

Before delving into the specifics of portfolio management tools for decentralized finance, it's critical to grasp their key characteristics. Here are some of the features that must be included in an asset management project.

Composability

The ability to work with a wide range of decentralized finance projects is provided by well-known portfolio management projects, allowing for a smooth decentralized finance experience.

False-anonymity

The portfolio management tools for decentralized finance can link with each other via a wallet address. Participants can choose to reveal their identities if they like to or stay anonymous.

Non-custodial

The finest decentralized finance portfolio management projects do not suggest that the underlying assets are no longer owned. The assets remain in the wallet that is currently in use.

Universal accessibility

The essential feature of decentralized finance portfolio management projects is their ability to be accessed anywhere, regardless of tax bracket or location.

Automation

Automation is a feature of an ever-increasing number of portfolio management tools. Consequently, they may do smooth collateralization, liquidation, and rebalances without requiring participant input.

Features of DeFi Portfolio Management

The following are among the most prominent features of fund management products:

- Non-custodial – The underlying fund ownership is never removed, and they appear to stay in the wallet that is being used.
- Composable – Most of the top fund Management projects integrate with a variety of DeFi projects, resulting in a complete DeFi experience.
- Automated – An array of fund Management tools are automated, allowing for smooth rebalancing and collateralization without user intervention. Fund management tools are available to everyone, irrespective of where they live or their tax bracket.
- Pseudo-anonymous – Fund Management products often

communicate through a wallet address, implying that sharing one's identity is optional.

What to Expect from a DeFi Portfolio Manager

Regrettably, not all decentralized finance portfolio managers are equally useful and safe. Though each portfolio manager will cater to a different sector of investors, all of the options you choose should include at least these three fundamental qualities.

- Security and safety

Decentralized finance and other cryptos are tough to manage and trace due to the decentralized structure of the blockchain network on which they operate. It's equally tough to figure out who owns each wallet, and it's very hard to reverse a transaction once it's been conducted.

This emphasizes the need to select a manager that prioritizes security. Since most DeFi portfolio managers connect to an exchange, it's critical to safeguard your tokens and funds. Security functions such as two-factor authentication and multi-layer encryption may protect the wallet from a single attack or data leak.

- A Simple and Straightforward Design

The goal of managing a portfolio is to examine all of your assets easily and monitor how their value fluctuates. On the first page of a great portfolio manager, or with only one or two clicks, you may see the overall worth of your portfolio.

Your manager ought to be straightforward to use and comprehend, and you shouldn't have to strain to figure out how much each token in your portfolio is worth. Consider a manager that has a simple, clean design that you want to look at and use. When utilizing your portfolio manager, will help you save considerable time.

- A Diverse Range of Tokens and Exchanges

A portfolio manager should be able to keep track of all of your tokens.

Look for a DeFi portfolio manager that connects to all of the tokens you own and all of the exchanges you use.

Carefully consider the system's currency offerings before signing up for an account if you're looking for a combined manager and exchange. This allows you to trade as many tokens as possible without dealing with numerous exchanges or remembering passwords.

DeFi Fund Management Projects

When we speak of decentralized finance portfolio management tools, digital wallets are the first thing that springs to mind. In addition, programs specializing in handling cryptocurrency asset portfolios are among the first things to consider when it comes to portfolio management tools. Below is a list of all the DeFi portfolio management tools.

- Balancer
- Curve
- DeFi Saver
- DeFi Score
- Inspect
- InstaDapp
- TokenSet Protocol
- Zapper
- Zerion

We'll focus on Token Sets to help you understand how fund management works in a decentralized ledger.

TokenSets

TokenSets is a cryptocurrency network that enables users to purchase Strategy Enabled Tokens (SET). These tokens feature automated asset management methods that simplify handling your crypto portfolio without having to conduct trading strategies manually. With an automated trading strategy, you won't have to manually track the market 24 hours a day, seven days a week, decreasing missed opportunities and risks from emotional trading.

Each Set is an ERC20 token made up of many cryptocurrencies that reshape its holdings depending on your strategy. To put it another way, SET uses tokens to execute crypto trading strategies.

Types of Sets

Sets are divided into two categories:

- I Robo Sets and
- Social Trading Sets.

Robo Sets

Robo Sets are automated trading strategies that trade tokens according to predefined criteria in smart contracts. There are four major categories of algorithmic strategies presently in use:

- Buy and Hold: This strategy reconfigures the portfolio to its target allocation to avoid overexposure to any single token and spread the risk across several tokens.
- Trend Trading: This strategy employs Technical Analysis indicators to move from the target asset to stablecoins following the strategy.
- Range-Bound: This strategy works in bearish or neutral markets and automates buying and selling within a set range.
- Inverse: Those who want to "short" a benchmark should use this strategy. When traders believe a model is due for a correction, they will buy this.

Social Trading Sets

Social Trading Sets allow users to follow the top trading strategies of a few featured traders on TokenSets. By purchasing this Social Trading Set, you will be able to copy the trades made by these featured traders readily. Social Trading Sets are algorithmic, but rather than being produced by the TokenSets team like Robo Sets. Well-known traders create them.

Benefits of Using Sets

Sets tokenize trading methods. Set is the simplest way to try out some of the chosen trading strategies or follow experienced traders' footsteps.

Having said that, always do your homework. Just because a Set has performed well in the past does not guarantee that it will continue to do so in the future. The crypto market is volatile, and the adage "past success is no guarantee of future results" is particularly true in this case. Rather, do some analysis and comparison searching to see which strategy makes the most sense for you, and then use TokenSets to get started quickly.

We'll use the ETH/BTC RSI Ratio Trading Set as an example of one of the highest-rated Robo Sets. In this case, the Robo Set uses the Relative Strength Index (RSI) technical indicator to implement the Trend Trading Strategy. The value of this trading strategy increased by 102.33 percent, compared to 41.29 percent for holding BTC and 94.17 percent for holding ETH. Since Token Set is so new, there is only output data for the last three months at writing.

That's all there is to TokenSets; if you want to get started or try it out, we've included a simple guide for having Sets.

Simple Guide to TokenSets

Step 1:

- Visit https://www.tokensets.com/ for more details.
- Select "Get Started"
- Click "Next" until your wallet receives a request.
- To connect your wallet, click "Connect."

Step 2:

- Accept the Terms of Service and the Privacy policy.
- Your email address is optional.

Step 3:

There are two types of Sets:

- Social Trading Sets
- Robo Sets

You have the option of selecting which sets you want to purchase.

Note: You can do your thorough research and analysis before purchasing any sets!

Step 4: We have selected Robo Sets.

- Select "ETH/BTC RSI Ratio Trading Set" from the drop-down menu.
- Click the "Buy" button.
- Input the number of sets you'd like to purchase.

Step 5:

- Before trading, you must enable Dai if you are a first-timer.
- You will proceed with your purchase once your approval has been verified.

Step 6:

Done!

Zapper

Zapper is steadily causing a stir as a sector-leading fund management tool, as it is the DeFi dashboard is presently monitoring the most DeFi goods. Zapper is well-positioned to handle and invest resources into top DeFi networks, thanks to its latest merger with DeFiSnap and DeFiZap.

Why Zapper?

- Keep an eye on top DeFi protocols like Uniswap liquidity pools and Synthetix, as well as emerging projects like Curve.

- Easily monitor any debt owed through DeFi items to see which positions can need supplementation.
- Using their Protocol Allocation method, you can see how your DeFi positions are distributed by percent.
- With ROI forecasts, you can monitor how different positions are likely to do.
- Pool Pipes are used to linking resources from various liquidity pools.

Zerion

Zerion has rapidly established itself as one of the sleekest fund management products as a single dashboard for quickly tracking position across an infinite range of DeFi products. Zerion's dashboard presently allows users to Save, Exchange, Invest and Borrow with a range of DeFi's most well-known brands, lending credence to the platform's monitoring capabilities.

Why Zerion?

- Zerion works with a variety of web 3.0 wallets and is non-custodial, which means it doesn't require users to deposit their funds into the network itself.
- A QR code can be used to connect a user's Zerion dashboard to their mobile device. This helps users keep track of their positions on their phone and desktop devices without ever needing to move funds from their wallets.
- Users can access Compound, Uniswap, Maker, and other tools from a single dashboard in only a few steps.
- Outside of the support products, Zerion keeps track of a user's Set Protocol positions and every token in their wallet.

DeFi Saver

DeFi Saver, as a rapidly growing fund management network, is the ideal place to control and retain outstanding lending and borrowing positions, especially those of Maker CDPs (Vaults). DeFi Saver, which was built with Flash Loan capabilities, provides a series of automation tools to help users properly manage their collateralization ratios while keeping failsafe in place in the event of large price spikes, such as those seen on Black Thursday.

Why DeFi Saver?

- Build, administer, and maintain a Maker Vault from a single location.
- Save funds with Compound, dYdX, Fulcrum, and the DSR.
- DeFi Saver allows people without technological experience to engage in Maker liquidations.

Gnosis

Many decentralized finance portfolio management projects are incapable of supporting portfolio management by teams. Gnosis fills the gap by functioning as a portfolio management tool designed specifically for users interested in managing cryptocurrency assets as team members. Individual traders will also benefit from Gnosis.

Gnosis is among the best portfolio management tools in decentralized finance since it provides many features. It allows for the storage, investment, and trading of cryptocurrency assets and unrestricted access to a variety of DeFi applications. Besides portfolio management, Gnosis additionally provides services for corporate users, like payroll administration.

dHEDGE

dHEDGE is another significant addition to the finest decentralized finance portfolio management tools. It's essentially a decentralized portfolio management platform with Synthetix inclusion. The tool enables users to connect with portfolio managers in a smooth manner. According to dHEDGE's promises, the Ethereum blockchain network can connect users with some of the world's best portfolio managers.

Transparency is emphasized throughout the protocol and related platforms as a prospective feature of decentralized asset management. Before investing in dHEDGE, investors may simply check the track records of asset managers in a transparent manner.

Most importantly, the dHEDGE platform leaderboard allowed users to track an asset manager's performance. It also comes with a 'Performance Mining' function that pays users who invest in lucrative pools on the platform. Holders of DHT tokens are accountable for resolving dHEDGE governance issues.

Yearn.finance

Yearn.finance might be the place to go if you're looking for the best DeFi portfolio management tools. It is one of the most widely used decentralized finance software packages. Vaults is a prominent asset management tool in the Yearn.finance portfolio that assists investors in investing their money.

Following that, the cash is invested using several unique techniques designed to maximize the income from the asset deposits. At the same time, the techniques are intended to alleviate asset management risk issues. The primary benefit of Vaults is the elimination of transaction costs associated with individual trading, which is made possible by pooling investors' assets.

6

DEFI OPTION
TRADING PROTOCOL

What is DeFi Option Trading

eFi options are borderless, low-barrier tools that may be exchanged among peers on the network without brokers or the payment of a commission. Options offer a trader the right but not the responsibility to purchase or sell an asset at a particular future date. Options are among the most prevalent forms of derivatives in the conventional financial system, and they are simply a contract between an option buyer and the seller. Similarly, it implies a responsibility for the option seller to purchase or sell the underlying asset from/to the buyer. An options contract gives the buyer the chance and choice to buy or sell an underlying asset at a pre-determined price (called the strike price). It's similar to a financial tool with particular conditions, like "if this, then that."

Basic Concepts of Option Trading

Calls and Puts

Calls and puts are the two forms of options. Each one denotes a particular action that will be guaranteed to occur in the long run if a certain condition is satisfied.

The buyer of a call option has the right to buy the concerned asset at the strike price, whereas the buyer of a put option can sell the concerned asset at the strike price.

American and European Options

The buyer's contract can be exercised in one of two ways: American or European. The buyer may use their right to purchase or sell until the contract expires with American options. However, the contract buyer could only use their right after the option's expiration date for European options.

Cash and Physical Settlement

An option could be exercised in one of two ways: cash or physical settlement. If both entities consent that just the counterparts' payout discrepancy will be offset, the cash settlement is made. The major result is that the holder of the asset would not change. Let's look at an example to see what I mean:

Jakob offers John an American put option with ETH as the underlying and a $1500 strike price. When the price of ETH falls below $1000, John chooses to execute the option. Rather than transferring 1 ETH to Jakob, John receives the strike price less the spot price (1500–1000), which equals $500.

As the name implies, the physical settlement occurs whenever the underlying asset in possession is transferred from one party to another after the option is executed. In this scenario, Jakob would have paid $1500 to John, and John would have transferred 1 ETH to Jakob, as in the earlier example.

Physical settlement is presently used in our decentralized finance options. This implies that anytime an individual wants to exercise their options, they must transfer the whole value of the underlying asset to the contract to collect the contract's collateral.

Premium

The premium is simply the market price of an option. It indicates the option seller's remuneration for writing and selling the contract. It may equally be thought of as the expense incurred by the option buyer in hedging his present position. The premium is paid in advance.

Strike Price

The strike price is an option contract element. It is simply the price at which the underlying asset would be transferred if the option is exercised.

For call options, the strike price is the amount at which the option buyer will pay to acquire the underlying asset from the option seller. Whereas for put options, the strike price is when the option buyer will offer the asset to the option seller.

Underlying Asset

This is the asset that is being bargained to be bought or sold when the option is exercised.

If we indicate that the underlying asset of an option is ETH, we mean that when the price falls short of the strike price and the buyer uses their right to sell the option, the ETH is the token offered to the option seller strike price.

A similar concept applies to call options, except that rather than being the asset sold, it would be an asset purchased at the strike price specified by the option buyer.

Collateral Asset

The collateral asset is simply the option seller's item in the contract to ensure that the transaction is honored once the option buyer decides to sell or purchase.

The put option in the instance of an ETH:USDC, the collateral asset will be the second asset since the option buyer will sell the ETH for the agreed-upon price in USDC. In call options on ETH:USDC, the collateral asset will be ETH, since the option buyer will exchange their USDC for the option seller's ETH.

When to Exercise an Option

Assume Amanda has purchased a European put option on ETH:USDC with a strike price of 1800 USDC, which will expire in a matter of hours. How can she determine if exercising the option will be advantageous for her or not?

An option might be in-the-money, out-of-the-money, or at-the-money in the market. Let's have a look at what they imply.

In-the-money

This is a circumstance in which the owner would profit by using the option. When it comes to put options, we may state that if the strike price is greater than the spot price (the token's market price), the option buyer will make a gain.

Condition for in-the-money options at expiration

That indicates that if she uses her option, she will sell the asset to the option seller at a greater price than the market, as agreed. Therefore, when the strike price of the ETH is presently 1000 USDC, Amanda should use her option in order to make an 800 USDC gain.

The result of exercising in-the-money options, per option exercised

Long Put Profit/Loss Graph — The risk is confined to the premium for put options, and the gain is constrained until the asset's spot price falls to zero.

Out-of-the-money

When executing an option is not advantageous for the option buyer, the option buyer may lose money. Let's look at Amanda's example again, but this time with the spot price at 2000 USDC. Why would she sell her ETH below the price market is willing to pay for it? If she executes the option in this instance, she will sell the ETH for 1800 USDC, losing a possible gain of 200 USDC.

At-the-money

This is the spot price, and the strike price is the same.

Condition for at-the-money options at expiration

In this situation, the option owner should primarily evaluate whether or not to maintain the underlying asset, which in Amanda's example is ETH.

Long Call Profit/Loss Graph — As the price of the asset rises, it is clear that the loss is confined to the premium, and the gain is infinite for call options.

Trading and Pricing

One of the most significant distinctions between traditional and decentralized finance options is exchanging and how they are constantly valued. Options are usually exchanged within orders books in traditional markets, and markets include a large system of market makers who continually price and place orders in various options markets.

The Automated Market Maker, or AMM, the decentralized finance landscape processes buy and sell orders to reposition order books. The system can ensure liquidity for trades at any moment by establishing liquidity pools for options and stablecoins. Transaction costs from the trading activity are paid to liquidity providers.

The Options AMM enables single-sided liquidity provision while pricing options programmatically using market circumstances and Black Scholes Model.

To generate a derivative price, general derivatives pricing formulas require one or more market variables (such as the spot price of the underlying asset). External factors are coupled with calculated characteristics (like time to expiration and risk-free rate) and internal elements (like implied volatility) in the Black Scholes model.

Black Scholes Model

A model for calculating the price of an option. Through the present trading price of the underlying asset, the strike price of an option, the time to expiration, the projected dividends, the projected interest rates, and the volatility, the Black-Scholes model may be used to calculate the value of an option. Although the Black-Scholes model is not perfect, it is nevertheless extensively employed to compute prices today. It's also used to determine an option's implied volatility.

Option Trading Protocol

The platforms for cryptocurrency options are different from those for traditional finance options. Decentralizing a protocol and putting all transactions on-chain comes with substantial technological challenges that protocol developers must overcome.

Nonetheless, several DeFi protocols have taken varied approaches to these issues, each with its own set of advantages and disadvantages. The following are the results of my investigation of these numerous projects:

Siren Markets

Siren Markets ('SIREN') is a decentralized Ethereum platform that allows users to create, trade, and fully-collateralized option contracts for all ERC20 coins. Siren is a fully collateralized options network with no need for oracles to settle trades. With both the long and short sides of the contract tokenized in this system, it is easy for options traders to enter and exit their holdings at any moment.

A bToken represents the buyer's side, whereas a wToken represents the writer's side. A secondary market for short and long exposures is enabled by tokenizing both sides of the contract. SIREN's pricing options are approximated using an on-chain Black-Scholes model.

Owners of bTokens can purchase (for calls) or sell (for puts) the underlying asset at a specific strike price. Owners of wTokens can redeem collateral if the option hasn't been executed or the payout from a matured option contract.

Providing liquidity to a SirenSwap AMM allows you to become a writer. When an option buyer buys a bToken via the SirenSwap AMM, this liquidity is combined with other liquidity sources and utilized to underwrite the option. The option buyer receives bToken, whereas the AMM pool receives wToken.

To exchange bTokens and wTokens, SirenSwap AMM employs a hybrid of a constant-product bonding curve and options minting. Except for bTokens and wTokens, which may be traded against the collateral token, the SirenSwap AMM does not need any assets in the pool.

SI is Siren Markets' governance token. Owners of the SI token will soon be able to partake in protocol decision-making, such as the development of options markets and reward schemes, among other things. Trading costs will accumulate to SI token owners whenever fees on the platform are activated.

Opyn

Opyn(v1) was the first decentralized finance options platform to leverage Uniswap's AMM liquidity and uses a 0x order book in v2.

The price in Opyn indicates the amount a premium user must spend to purchase one oToken of puts and calls. Opyn v1 is a two-way platform where you may purchase or sell options, not only sell to close out for intrinsic value, and where option pricing is decided by market forces (0x).

It's all about impermanent loss vs. fee revenue without the option. However, with the option, it's about impermanent loss, which can be hedged. Thus it's option-cost vs. fee-income. In comparison to unexpected price fluctuations compared to the temporary loss, fee income is reasonably predictable by checking pool usage on Opyn and evaluating trade volume/liquidity patterns.

The option's cost (time-price) is a negative offset. Therefore the baseline begins with a negative value. You try to make money through LP-ing in

Opyn. It's an excellent bargain if the charge exceeds the cost of the option.

Opyn v1 was completely collateralized in a strike for puts and an underlying asset for calls, allowing naked (uncapped) options to be written and acquired. For the 'long tail' of options, physically settled options will be available in v1.

Lower margin requirements are available in Opyn v2, with spreads requiring a maximum loss and naked options requiring less than a maximum loss at first. Additionally, sellers would be able to receive yield and governance collateral token earnings. Furthermore, v2 adds governance features for future decentralization.

Hegic

Hegic is a cash-settled, American-fashion options platform that links option buyers and sellers using a peer-to-pool options architecture. Since it continuously combines liquidity from all players in the market, pooled liquidity is fantastic. Hegic sellers must offer liquidity to the Hegic pool when they sell their call or put options for a premium. As a result, Hegic's liquidity Providers are option sellers, and the pool offers collateral for a variety of options at once. Anybody interested in becoming a liquidity provider (LP) at Hegic may make a deposit and take on their pro-rata portion of the total seller liquidity. Additionally, only after the option is bought is it written down.

Buyers may choose their underlying asset, duration, and strike price using Hegic. The prices are calculated using a program depending on the contract features that have been selected. A Black-Scholes approximation is used to calculate the price of an option, and skew.com is used to include implied volatility (IV) into option prices. Hegic, on the other hand, does not make any extra price adjustment for market fluctuations like supply and demand, which implies that a Hegic pool might run out of collateral if the underlying asset price alters dramatically. Hegic, on the other hand, has a pooled LP strategy that allows it to sell options to purchasers at any strike. Hegic pools overprice non-ATM options, leaving purchasers with a terrible deal; nevertheless, any AMM that uses the standard Black-Scholes approximation will do the same.

HEGIC is Hegic's own native coin. The HEGIC coin is eligible for a portion of the protocol transaction cost (1 percent of all options volume traded on Hegic). Hegic transaction costs are paid in the associated asset of the option, which implies that takers get the underlying asset's return.

Potion

cPotion is a newbie to the cryptocurrency world. It's a one-sided option AMM that protects against temporary losses caused by the cryptocurrency market's extreme price volatility. To put it another way, it acts as insurance for the position, reducing the risk of severe losses while maintaining the upside potential. However, it is important to note that this "insurance" does not imply that the owners will not suffer losses in a dynamic situation, but rather the likelihood of such significant losses would be reduced with time.

Potion allows purchasers to select their underlying asset, duration, and strike price, after which they are offered a programmatically calculated pricing depending on those input factors. The Potion AMM is a bonding curve that is spread across several LPs. The team utilizes Kelly Criterion, which turns the order size into a bet size, to relate usage to premium. The system then advises on the proper premium size for the bet size and the suitable odds. The information is then plotted on a bonding curve.

Potion was founded by technological experts who believed that LPs were taking on more risk than they realized. They aim to account for supply and demand in their AMM, which employs the reverse Kelly Criterion. Potion began by focusing on LP survival rates to assure LPs that they will not go bankrupt at any point in the future. Potion aims to give end-users a wide range of alternatives by allowing them to personalize their selections and choose any strike and expiration time.

FinNexus

FinNexus provides its participants with a lot of freedom and options. It's a peer-to-pool options platform that provides LPs with pooled liquidity and reduces risks. Therefore growing liquidity is a primary concern. When it comes to option pricing, FinNexus uses the Black-Scholes model. Additionally, the volatility surface is standardized for various strikes and maturities, and IV is their most significant metric that is supplied into their protocol.

FinNexus features two Ethereum pools, one using FNX and the other with USDC/USDT as collateral. The linked collateral asset is used to acquire and settle options. The MASP, which represents Multi-Asset-Single Pool, is another FinNexus service that enables hybrid assets to be used as collateral. MASP allows for collateral such as a steady coin pool of USDC/USDT. MASP enables the inclusion of any form of asset. MASP settlement is similar to traditional trade in that it's done in stablecoins. FinNexus also has mining and staking services.

FinNexus has a wide range of underlying assets to choose from. As a result of the diversity of underlying assets coupled with the in-house risk adjustment criteria that function as an AMM process and stabilize the put and call distribution within the pool, LPs' risks are more spread. FinNexus Protocol for Options (FPO) is another FinNexus product. FPO, equally known as The Universal Options Platform, is a cross-chain permission-less system for options that have debuted on Binance Smart Chain, Wanchain, and Ethereum.

Ribbon Finance

Ribbon Finance has entered the decentralized finance Options market with a new financial-oriented product. With their Vault Products, they provide significant returns on ETH using an algorithmic option method. Opyn's oTokens are used in their option-writing techniques.

When compared to individual option transactions, this makes it simpler while still allowing for potential profits. To further decentralize their protocol, they just introduced Ribbon Governance ($RBN).

Charm

Charm aims to increase the basic options layer by simplifying the option pools and summing options to make it easier to estimate values. Charm's pricing is determined by market forces, which allows for price discovery. It equally features a unique AMM architecture that allows you to trade several strikes, although it's not perfect when continuously rolled after the expiration.

Charm is developing a vault where LPs may put money, and the funds will be immediately transferred to the next pool. The risks for a passive LP, according to Charms, are greater since there are more strikes, making hedging operations more difficult. Similarly, trading volumes are substantially greater since more options are accessible and slippage is lower due to the AMM architecture. Other pools (such as Primitive) are secure, but their returns are significantly lower due to a lack of liquidity. As a result, there is frequently a trade-off between LP investment and risk.

Lastly, the Charm fee is just 1%, and there are no further costs for closing or settling a position.

Primitive

Primitive features two ERC20 tokens: a LONG and a SHORT option. To make options, 100 percent collateralization is required. When a new option is formed, the Long and Short tokens are issued that may be sold to acquire exposure to the option. Uniswap adds liquidity to the system, which is used for option token transactions.

Opium

Opium is a protocol for both options and other derivatives, and it provides a range of option products. COMP, YFI, NXM, BAL, and other underliers have all been used to create goods. In addition, the Opium market allows participants to trade Opium products and offers liquidity.

Opium, in partnership with Aave, blazed the way for Credit Default Swaps (CDS), which insures against a borrower's default. It has created a CDS

Contract for a price decline of USDT/USDC and WBTC/BTC. Opium also introduced the first IRS product, in addition to CDS.

Hedget

Hedget is a P2P decentralized options platform with complete collateralization that enables European fashion options to settle only upon expiry. The HGET platform token is the governance and utility token for Hedget and is released via an ERC-20 contract on the Ethereum platform. Hedget is designed for both Binance Smart Chain and Ethereum, with Chromia serving as a Layer 2 improvement to the Ethereum network.

ACO

Auctus Options (ACO) is a non-custodial decentralized options system that is tokenized with order book liquidity. It establishes liquidity pools for sellers, and by selling covered options, these sellers earn premiums routinely. ACO intends to establish bespoke liquidity pools for its customers that select a minimum IV modified according to the pool's use rate.

The flash exercise function of ACO is an intriguing tool. Participants can employ Uniswap V2 flash swaps to offset the cost of executing their ACO options, leaving them with substantial gains. Flash exercise enables traders that wish to acquire exposure with less money to exercise lucrative options contracts in one transaction, even if they don't have the cash to purchase or sell the asset before maturity.

Finally, ACO made vault strategies available to its users.

Option Trading Strategies

Long call

The user buys a call – known as "going long" a call — expecting that the stock price will surpass the strike price by expiration. If the stock surges, the potential on this strategy is unlimited, and users can profit several times their actual investment.

For instance: If stock Y is selling at $20 for a share, a call with a strike price of $20 and a four-month expiration is trading for $1. The contract costs $100, equal to one contract * $1 * 100 shares per contract.

Covered call

A covered call is similar to trading a call option, although with a twist. In this case, the user sells a call while also purchasing 100 shares of the stock underlying the option. You may transform a potentially hazardous transaction – a short call — into a reasonably safe and profitable one by owning the stock. At expiration, users expect the stock price to be lower than the strike price. The holder should sell the shares to the call buyer at the strike price when the stock ends above the strike price.

For instance: If stock Y is selling at $20 per share, a call with a strike price of $20 and a four-month expiration is selling at $1. The contract offers a $100 premium, equal to a contract * $1 * 100 shares per contract. For $2,000, the user purchases 100 shares of stock and sells a call for $100.

Long put

The user acquires a put – known as "going long" a put — with the expectation that the stock price will be short of the strike price by expiration. If the price drops considerably, the profit on this transaction might be many times the initial investment.

For instance, if stock Y is selling at $20 per share, a put with a strike price of $20 and a four-month expiration is currently selling at $1. The contract costs $100, equal to one contract * $1 * 100 shares per contract.

Short put

This technique is the inverse of the long put, in which the user sells a put — known as "going short" a put — with the expectation that the stock price will be higher than the strike price by expiration. The user obtains a premium in return for selling a put, which is the maximum a short put may generate. The user must buy the shares at the strike price when the stock falls short of the strike price at option expiration.

For instance, if stock Y is selling at $20 per share, a put with a strike price of $20 and a four-month expiration is currently selling at $1. The contract pays a $100 premium, equal to one contract * $1 * 100 shares per contract.

Married put

This technique is similar to the long put but with a slight difference. The user acquires a put and holds the underlying stock. This is a hedging transaction, where the user expects the stock to increase but needs "insurance" in case it drops. When the stock does drop, the long put will compensate for the loss.

For instance, if stock Y is selling at $20 per share, a put with a strike price of $20 and a four-month expiration is currently selling at $1. The contract costs $100, equal to one contract * $1 * 100 shares per contract. The user spends $2,000 on 100 shares of stock and $100 on a put.

7

FLASH LOANS

What are Flash Loans

In DeFi landscape, flash loans are stealing the show, with all decentralized app scrambling to find out how to include this innovative function into their platform.

It is widely agreed that the decentralized finance world is here to stay. Decentralized finance protocols are constantly importing most traditional financial institution ideas into a decentralized landscape of smart contracts. On the other hand, flash loans take an entirely different approach since they are non-collateralized loans that are almost absent in traditional financial institutions, let alone in decentralized finance.

Now, let's take a closer look at flash loans and their various features.

Meaning of Flash loans

Simply defined, a flash loan in the decentralized finance landscape is the same as an unprotected or un-collateralized loan. In the traditional financial system, borrowers must present collateral as security for the loan, ensuring that the borrower will repay the debt. After reviewing the borrower's credit record and evaluating the collateral supplied by the borrower, the lender takes time to authorize the loan.

Flash loans, nevertheless, provide a novel idea in the decentralized finance landscape: a borrower may quickly borrow Ethereum's ETH or other ERC20 coins to reap the benefits of arbitrage possibilities in the DeFi landscape without having to put up any collateral.

The whole procedure is powered by smart contracts, which are meant to run arbitrary code when the borrower receives the loan, guaranteeing that the loan is financed in the same transaction. A flash loan's architecture may appear simple at first glance, but it necessitates complicated and advanced programs that enable developers and cryptocurrency investors to profit from arbitrage possibilities on DEXs.

Unique Feature of Flash Loan

- Supported by Smart Contract Rule

A smart contract is a piece of blockchain technology that contains all of the rules needed to conduct flash loan transactions. The smart contract guarantees that no transaction takes place until the borrower has repaid the loan in full before the transaction is finalized.

What Happens If the Borrower Fails to Pay?

Whenever the borrower fails to pay, the smart contract cancels the transaction, implying that the loan was never made.

- It's Quick and Easy

As the name implies, Flash implies anything that appears and disappears in a flash, instantaneous. Likewise, a flash loan is quick; the borrower must utilize smart contracts to execute instant transactions against the loan provided by the lender; this trade must occur before the deal and generally lasts a few seconds.

- It is Free of Collateral

As previously stated, a flash loan is an unsecured loan in which the borrower is not required to provide any collateral to obtain a loan. Having mentioned that it is collateral-free, we must realize that this does not imply that the lender will not be reimbursed for the money given.

It is possible through the usage of a smart contract. Rather than putting up security, the borrower is required to repay the money immediately before the contract ends.

Traditional Loan Vs. Flash Loan

Flash Loan

The majority of us are familiar with how a traditional loan works. Even so, it's worth repeating so that we can compare the two afterwards.

Unsecured Loans

An unsecured loan does not require you to put up any collateral. It is a loan where you don't provide the lender with any assets if you don't return the loan amount. For instance, let's say you want a $4,000 gold necklace with the Binance emblem on it. You don't have the money right now, but you will when you get your salary next week.

You have a conversation with your pal Bob. You tell him how much you want to get this chain and enhance your current trading game by at least 15%, and he decides to lend you the funds, obviously, on the premise that you return him when you get your salary.

Because Bob is a wonderful buddy, he didn't charge you interest when he loaned you the $4,000. Not everyone will be as generous – but why should they be? Bob has faith in your ability to repay him. Another individual may not be familiar with you, so they are unsure if you will steal their money.

Unsecured loans from financial institutions usually need some form of credit check. To determine your capacity to repay, they'll go over your credit history. They could assume you're fairly trustworthy if they notice you've taken out numerous loans and paid them back timely. Let's make a loan to them.

The financial institution then provides you with the money, but there are conditions attached. The conditions in question are interest rates. To receive the money now, you must accept that you will have to repay a larger sum later.

If you have a credit card, you may be acquainted with this concept. If you don't pay your bill for a certain amount of time, you'll be expected to pay interest until you pay off the entire sum.

Secured Loans

A decent credit score isn't always sufficient. Even if you've paid off all of the loans as at when expected for decades, you'll find it difficult to borrow substantial quantities of money merely based on your creditworthiness. You'll have to put up collateral in these situations.

When you ask somebody for a large loan, they may be hesitant to accept it. They'll insist that you place some money on the table to reduce their risk. If you don't pay back a loan on time, an item of yours - anything from a necklace to property – will be taken over by the lender. The assumption is that the lender will be able to recoup part of the money they have lost. That is collateral.

Assume you now desire a $50,000 automobile. Bob believes in you, but he refuses to offer you the cash as an unsecured loan. Instead, he requests that you put up significant collateral in the form of an item. Bob can now take and sell the item if you fail to return the money.

How it Works

A flash loan is an unsecured loan since you don't supply collateral. You do not, however, have to undergo a credit check or anything of the sort. You just ask whether you may borrow $50,000 in ETH from the lender, and they say yes! You've got it! After that, you're free to go.

What's the catch? A flash loan has to be paid back in one transaction. That doesn't seem particularly logical, but that's just because we're familiar with a transaction structure in which money is transferred from one user to another, much like when you pay for products or services or when you put coins in an exchange.

If you understand Ethereum, you'll know that it's a very versatile system, which is why some people refer to it as "programmable money." You may consider your transaction "program" for a flash loan consisting of three components: receiving the loan, doing something with the money, and repaying the lender. And it all unfolds in the blink of an eye!

Let's simply put it down to the blockchain network's brilliance. The transaction is sent to the blockchain, and the funds are temporarily lent to you. In the second part of the transaction, you may do different things. Whatever you choose to do, just make sure the funds are returned in time for part three. If they aren't, the system will deny the transaction, and the lender will receive their money back. Basically, they've always held the funds in terms of the blockchain network.

That indicates why you don't have to provide any collateral to the lender. The repayment contract is governed by law.

Takeaway

You're maybe thinking, why do you have to take out a flash loan at this point. You can't actually buy a Lambo if all of this happens in one transaction, can you?

That isn't the objective here, though. Let's look at the second stage of the transaction when you really do something with the money. The aim is to put the money in a smart contract, make a profit, and then return the funds to the original lender after the transaction. The aim of flash loans, obviously, is to earn.

There are a few scenarios in which this may be useful. In the meantime, you can't do anything off-chain, but you may use decentralized finance protocols to generate extra money with the loan. Arbitrage is one of the most common uses when you benefit from price differences between different trading sites.

Assume that a coin costs $10 on decentralized exchange A but $10.50 on decentralized exchange B. Buying ten coins on DEX A and selling them on DEX B will provide a $5 provided there are no costs. This type of transaction isn't going to get you a private jet immediately. However, you can imagine how trading a huge amount of money may help you earn extra cash. If you bought 10,000 coins for $100,000 and sold them for $105,000, you would earn $5,000.

You can benefit from arbitrage possibilities like these on DEXs if you get a flash loan (through the Aave protocol, for example). Below is an example of how that could work:

- Obtain a $10,000 loan
- Use the loan to purchase DEX A tokens
- On DEX B.
- Pay back the loan (plus any interest)
- Retain the profit

Everything is done in one transaction! In reality, however, transaction costs, along with heavy competition, interest rates, and slippage, make arbitrage profits razor-thin. To make the transaction lucrative, you'd have to figure out how to take advantage of pricing disparities. You won't have much luck competing against thousands of other people attempting to accomplish the same thing.

Flash Loan Attacks

Crypto, and by implication, decentralized finance, is an industry that is still in its infancy. When there are so many funds on the line, it's just a matter of time until flaws are uncovered. With the infamous 2017 DAO attack on Ethereum, we witnessed an illustration of this. Since then, 51 percent of protocols have been targeted for financial benefit.

In 2020, two high-profile flash loan attacks resulted in the theft of over $1,000,000 worth of the token. Both attacks had a similar pattern.

Some Famous Flash Loan Attacks

- dydx Flash loan attack:

In 2020, an attacker borrowed Ethereum ether flash loan from the dYdX lending decentralized application, split the loan fund into two halves, and sold it on two separate lending protocols, Compound and Fulcrum.

The attacker utilized a part of an Ethereum loan to short it against WBTC on the Fulcrum exchange, causing Fulcrum to purchase WBTC by passing this information to another decentralized finance protocol called Kyber, which then completed the transaction request via Uniswap decentralized exchange. Because Uniswap's liquidity pool had relatively little liquidity at the time, the price of WBTC skyrocketed, implying that Fulcrum overpaid for the WBTC it bought.

The attacker utilized the remainder of the dYdX loan fund during the Fulcrum trade to borrow WBTC as a flash loan from the Compound decentralized finance platform. Immediately the price rose, the attacker decided to flip the borrowed WBTC on Uniswap and earned a decent profit, then used the same to repay the dYdX loan and pilfered the leftover ETH.

This issue demonstrates the obvious weakness of the smart contract protocols, namely the bZx protocol utilized by Fulcrum. The attacker successfully fooled the bZx protocol into believing that WBTC was valued more than it truly was by participating in numerous transactions using five distinct decentralized finance protocols. This is simply market manipulation, with no genuine violation of the smart contract requirement of repaying the loan before the flash loan transaction was completed.

- PancakeBunny Attack

This is a recent occurrence that happened on PancakeBunny, a DeFi-based YF platform, in May 2021. PancakeSwap was utilized by the borrower (attacker) to buy a significant amount of BNB coins and influence the price of USDT/BNB and BUNNY/BNB in PancakeBunny's liquidity pools.

This allowed the attacker to amass a significant amount of BUNNY coin, which was then abandoned on the market to cause a price fall. Through PancakeSwap, he was able to pay off his loan. According to bscscan statistics, the attacker used the flash loan smart contract protocol to profit over $3 million.

In reality, in 2020 and 2021, there were many more flash loan attacks, highlighting the existing weaknesses in the lending systems that allow flash loans as a business.

How to Prevent Flash Loans Attack

Because attacks rely on decentralized exchanges recognizing their own or a single price feed that may be altered by making a large order for a cryptocurrency, it's a good idea to use decentralized pricing oracles to determine an asset's true price.

A dApp may defend itself from flash loan attacks in a variety of ways, the most popular of which are:

- Decentralized Oracles

Using decentralized oracles that use many sources to determine the 'real price' is by far the best option. Certain decentralized oracles, like Umbrella Network, put in extra effort to commit data to the Blockchain network to verify its authenticity.

This implies that if a malicious party attempts to launch a flash attack on a decentralized Oracle-based Dapp, the price alteration will fail, the transaction time will pass, and the whole transaction will reverse.

- High-Frequency Pricing Updates

On paper, this seems like a straightforward remedy, but it might be more costly in reality. It's simply about raising the number of times the liquidity pool asks an oracle for a new price. The reasoning is that when more changes are made, the price of a coin in the pool will be changed more quickly, invalidating the price alteration.

- Time Weighted Average Pricing

Traditionally, the average (or the median) has been used to determine the price in the liquidity pool. TWAP, on the other hand, recommends average prices across many blocks.

This mitigates flash loan attacks because the full series of attacking transactions must be executed inside the same block. However, the TWAP cannot be altered without compromising the whole Blockchain network.

Another effective method of preventing these attacks is to utilize two transaction blocks instead of a single block for the transaction cycle. It's likely to make the procedure more difficult and dissuasive for the attacker. Nevertheless, it carries the risk of negatively impacting the decentralized finance UI.

Certain protocols also incorporate tools for detecting flash loan attacks, which aid in early detection, quick reaction, and neutralization. Nevertheless, unless there are many examples of prevented attacks, it is difficult to prove these techniques' success.

Flash Loan Applications

Flash Loans can be utilized for legal reasons like arbitrage, liquidation, and so on. Furthermore, aggressive users may use Flash Loan as a potential weapon to destroy the decentralized finance landscape. This section will go through four different uses for Flash Loan and how they can profit from it.

Arbitrage

Arbitrage in decentralized finance, in general, is a strategy for gaining advantages by trading across platforms that offer different prices for the same item. Because the decentralized finance market responds slower to network events than the real market, investors can benefit from market inefficiencies to purchase and sell cryptocurrency assets at various prices to profit financially. It's worth noting that arbitrage isn't necessarily a bad thing. In reality, arbitrage can be used to bring decentralized exchange token prices closer together.

Benefit

Investors can start arbitrage with Flash Loan even if they don't have any pre-owned assets. If a pricing gap is discovered, arbitrageurs can use the Flash Loan service to borrow a large asset quickly and gain rewards. As a result, arbitrages with Flash Loan become "cost-free" if investors can pay the transactions.

Example

We shall discuss a four-step arbitration that occurred on January 22, 2021. First, the trader borrowed 1.13 Ether from dYdX in transaction 10. Second, a deal in Balancer turned 1.13 Ether to 345 LPT coins. Third, a deal was initiated to trade 345 LPT coins on Uniswap's matching liquidity pool. Consequently, the trader made a profit of 1.46 Ether. Ultimately, after earning 1.13 Ether, the trader remitted it to dYdX. The trader earned 0.33 Ether (approximately 538 USD at the time) by paying 0.05 Ether as a transaction cost.

Wash Trading

In a decentralized finance landscape, wash trading is a practice that generates fictitious trading activity for specific cryptocurrency assets or platforms. Wash trading is a collection of trades that increase the trading volume on a particular asset or protocol. In an actual sense, Wash trading has the potential to deceive users into doing financial transactions on the targeted cryptocurrency assets and platforms.

Although several nations, such as the United States, have prohibited washing trading from safeguarding their traditional markets and the stock markets, the appeal of crypto and the absence of regulatory oversight have brought it back to the cryptocurrency market.

Benefit

Wash traders can alter the market without a huge quantity of funds to bear the risk of loss and the transaction cost with Flash Loan.

Example

A trader borrowed 10 Ether from Aave protocol in transaction 11 on the 14th of July, 2020. Then, to improve the trading volume in the liquidity pool "Uniswap V2: DAI 2," five additional transactions were initiated. In specifics, 2 Ether was changed to DAI for each transaction, and all DAI was quickly changed back to Ether at the pool. After then, the trader repaid the Flash Loan. Other than wash trading, there were no other activities in this trade. Consequently, the user lost 0.068 Ether and paid 0.164 Ether as a transaction cost in this trade.

Flash Liquidation

Liquidation is the process whereby the liquidator purchases undercollateralized assets from lending protocols. There are two types of liquidation (Fixed Price Bidding and Auction), each with three functions (platforms, liquidators, and collateral holders). Liquidators can purchase undercollateralized assets from collateral holders with a predetermined discount using lending protocols such as dYdX and Compound for fixed price bidding. Furthermore, collateral holders will be subject to a specific liquidation fine imposed by the lending protocols. Conversely, protocols such as MakerDAO enable liquidators to vie like an auction on the holder's undercollateralized assets.

The winners who pay a greater transaction cost to start their trades reduce the undercollateralized collateral.

Benefit

Anybody can be a liquidator with Flash Loan and make money without a lot of money by purchasing undercollateralized assets at a certain discount.

Example

On November 3rd, 2020, transaction 12 completed a liquidation. The liquidator took a loan of 12, 940 DAI from dYdX and exchanged it for 13, 046 USDT. Second, 13, 046 USDT was utilized to purchase the asset from

Compound's undercollateralized holding. The liquidator received 13, 450 DAI in exchange for the purchased asset. Finally, after repaying the Flash Loan, gains of 510 DAI (about 510 USD) left, larger than the transaction cost of about 172 USD.

Collateral Swap

In decentralized finance landscape, a collateral exchange is an established behavior that involves two basic steps:

- Swapping: Getting rid of the former loan's collateral.
- Operation: Laundering operations using redeemed collateral.

Because the cryptocurrency market is so unpredictable, holders must close existing collateral positions as soon as possible to avoid losing money due to catastrophic slippages and liquidations.

Benefit

Without adequate money for Swapping, traders can take advantage of Flash Loan's "cost-free" assets to protect their collaterals from price slippage and liquidation. Furthermore, Flash Loan allows for Swapping and Operating activities to be combined into a single transaction. It also protects consumers from transaction-to-transaction uncertainty (such as slippage).

Example

A trader did a collateral swap to take out BAT (Swapping) and trade it for a more secure token USDC (Operating) in transaction 13 initiated on March 17th, 2020. The trader had previously established a DAI loan by depositing the asset BAT. To begin, the trader took out a 25 DAI Flash Loan from Aave. Second, 504 BAT were redeemed for 25 DAI (collateral).

Finally, the redeemed collateral was exchanged for USDC, a more secure token. Finally, the trader used some of the USDC to repay Aave's Flash Loan. Consequently, the trader's collateral was switched to a more secure asset, and they no longer held any DAI.

Finally, in the decentralized finance ecosystem, Flash Loan offers a great deal of flexibility for a variety of uses (arbitrage, wash trading, liquidation, and collateral swap). It has the potential to profit users or to be used by an attacker. As a result, both users and developers must be aware of the application's purpose.

How to Make a Flash Loan Using Aave

Aave was the first to propose the concept of a Flash Loan in the decentralized finance landscape. Prior to introducing flash loans, you had to put up an over-collateralized asset as collateral to borrow another one. If I planned to borrow a DAI, for instance, I'd have to put another coin with a higher value. Flash Loans shattered this notion.

All Ethereum transactions have a characteristic that allows for the creation of Flash Loans. And atomicity is an important aspect.

An atomic transaction is a sequence of indivisible and irreducible processes in a transaction. The Flash Loan takes advantage of atomicity to enable users to borrow money without putting up any collateral.

Remix Setup

We'll use the Remix IDE for the purpose of clarification.

The ability to build, debug, deploy, and alter Ethereum Smart Contracts is included in Remix.

Because the idea of this section is to describe how to make Flash Loan, we won't go into much detail into the IDE. Nevertheless, you need to familiarize yourself with the four key parts: main panel, side panel, icon panel, and terminal.

You need to download a browser extension that will enable you to interact with the Ethereum network before constructing the smart contracts. Various tools enable this feature, but MetaMask is the most common one.

MetaMask Setup

Detailed guide on how to install MetaMask.

- To begin, go to the website and download the extension.
- Accept the policy agreement by clicking on your freshly installed extension.
- Create a strong password!
- Make a mental note of the mnemonic seed phrase. This is something that should be preserved in the real world rather than on your computer.

If you've followed the four procedures listed above, you're set to begin drafting your first smart contract!

The Smart Contract

The smart contracts enable you to read and write data to the blockchain network by running deterministic algorithms. You use the Solidity programming language when creating a smart contract for the Ethereum network. Solidity files end in the.sol extension.

When you initially start Remix, you may remove any files that could be in the workspace.

You'll need to create a few different files:

- FlashLoan.sol
- FlashLoanReceiverBase.sol
- ILendingPoolAddressesProvider.sol
- IFlashLoanReceiver.sol
- ILendingPool.sol
- Withdrawable

FlashLoan.sol is implemented in the code snippet below.

This Flash Loan will borrow 1 DAI

To simplify, you begin by importing the dependencies necessary to run the Flash Loan. Abstract contracts are among the dependencies. Usually, a function in an abstract contract won't be implemented. You may consider it as a house design. A contractor uses this design to build a house. In our example, the design represents an abstract contract, you represent the contractor, and the house represents the derived contract.

In this case, the Flash Loan contract uses an abstract contract referred to as FlashLoanReceiverBaseV1. It includes all of the essential implementation information, like how to pay back the Flash Loan.

Now we'll have a look at the code line by line.

- First, we must specify the version of the Solidity compiler. Let's assume 0.6.6 in our context.
- Import the smart contract's dependencies
- The FlashLoanV1 contract is coined from the FlashLoanReceiverBaseV1.
- You pass an Aave's Lending Pool Providers address. In this example, we're giving the DAI Lending Pool's address.
- You develop a function called flashLoan. We give it the address of the coin we'd like to flash loan. DAI is the coin in this scenario.
- Because we don't require any data for the flash loan, we'll pass an empty string.
- We determine the quantity of DAI we would like to loan (in respect to wei, which is 10^18).
- To call the flashLoan function, you initialize the LendingPool platform, which is ILendingPoolV1 given by Aave.
- Lastly, the flashLoan function is called. There are four primary parameters to the function. First, you provide the address to which the loan will be sent. It's the contract in our own case. Second, you provide the token's address. In this example, it's the Kovan platform's DAI address. Third, you pass the asset amount, which is 1 "ether" amount of units (or 10^18 in "wei" units) in our example. And finally, you pass the data value, which is an empty string in this example.

The second function, executeOperation, is then defined. It's here that the flash loan comes in handy. After the flashLoan function has been fully completed, it is invoked internally. It requires four main parameters: -

The reserve address to which we will be required to repay the loan amount.

The value of the token

The transaction cost that the protocol charges.

An extra parameter that the function uses internally.

- It checks to see if we got the correct loan amount; otherwise, an error notification is displayed.
- In this stage, you have to implement the logic for any arbitrary use case.
- Through the SafeMaths library add function, we sum the fee and the loan amount.
- Finally, we repay the lending pool the complete debt or loan amount.

Deployment of the Contract

- To begin, open MetaMask and select "Kovan Test Network" as your network.
- Define the dependencies for the flashloan smart contracts using this gist. Copy the code from all the links and paste it into the Solidity file you created previously.

ILendingPool

IFlashLoanReceiver

ILendingPoolAddressesProvider

FlashLoanReceiverBase

Withdrawable

- Move to "Solidity Compiler" tab. Set the compiler to 0.6.6 click "Compile FlashLoan.sol"
- There should be a few warnings but no errors.
- You're now ready to launch the contract on the Kovan platform. Move to "Deploy & Run Transactions" tab. In the environment field, switch to Injected Web3 from JavaScript VM. This would prompt MetaMask to ask for your consent.
- Ensure that the FlashLoan.sol is selected in the "CONTRACT" section. Under the text field close to the deploy option, type in the LendingPool address. It will be 0x506B0B2CF20FAA8f38a4E2B524EE43e1f4458Cc5 in our case. Then select "Deploy." This will launch MetaMask should be launched.
- Select "Confirm" from the drop-down menu. You will receive a success message from MetaMask afterward. In the side panel, there will now be a "Deployed Contracts."

Funding the Flash Loan

You can copy the deployed contract's address from the new "Deployed Contracts" tab. We'll return to this later; for now, we'll have to update our Flash Loan contract with some DAI. This is the case since Flash Loans require money in the contract to complete.

You'll add the DAI token to MetaMask after you've received confirmation. To do so, launch MetaMask. Close to the bottom of the page, click "Add Token."Enter 0xFf795577d9AC8bD7D90Ee22b6C1703490b6512FD in the "Token Contract Address" column. This is the ideal address for DAI in Kovan. On hitting the "Next" button, It should show the DAI you received from the faucet previously.

After that, select the DAI token. When you click "Send," a window will appear. Enter the contract address for the Flash Loan, which we learned about earlier. Fill in the amount you plan to send. In our analogy, you will send 10 DAI. Then select "Next." Select "Confirm" from the drop-down menu. The Flash Loan contract 10 DAI would be successfully given.

Executing the Flash Loan

Return to Remix if you haven't already. There is a new "flashloan" text column in the deployed Flash Loan contract. The contract address of the token you plan to use is entered in this section. In our context, the DAI contract for the Kovan Testnet is 0xFf795577d9AC8bD7D90Ee22b6C1703490b6512FD. After you've entered the column accurately, click the "transact" button as seen below.

When you click the button, MetaMask will appear and ask you to approve the transaction. After you approve the transaction, you will receive a success notification. You should notice a URL in Remix's terminal. You must be routed to Etherscan after clicking that.

You will see three separate transactions under "Tokens Transferred."

- The red arrow indicates the transfer of 1 DAI from LendingPool to our contract.
- The repayment of 1 DAI and the associated transaction cost back to the Landing pool are indicated by the orange arrow.
- The interest accrued by DAI is indicated by the blue arrow, which has its own utility.

CONCLUSION

Decentralized finance (DeFi) is a network of financial apps built on the Ethereum blockchain networks. The apps that make up this DeFi industry are built on permissionless networks, allowing users to engage in a variety of common financial transactions without relying on a trusted central authority. Among these products are borrowing and lending, derivative exchanges, derivative, and decentralized insurance.

Banks and other financial organizations allow lending with deposits from other users in conventional finance. The borrowers' interest payments are used to incentivize lenders. It works in the same way in decentralized finance. However, DeFi lending and borrowing incorporates a trusted intermediary. Aave and Compound are two popular decentralized finance lending protocols we previously explained.

Currency exchange is another central aspect of the old financial system that is reflected in decentralized finance. In DeFi, users can swap one coin for another without the need for a central authority to keep track of their assets. As we have explained in a chapter, decentralized exchange platforms are divided into two types, Automated market makers and Order books. Derivatives, another important product in the fast-growing DeFi space, are agreements between two or more sides that specify how one side must pay the other. Options, futures, and swaps are types of derivatives.

There are numerous chances to benefit from DeFi as a completely new area in the cryptocurrency market. Indeed, DeFi's passive income-generating possibilities via yield-farming techniques fueled the sector's recent development. Decentralized finance, nevertheless, comes with several risks. The good news is that DeFi's disruptive potential makes it a sector ripe for growth. With our beginner to advanced guide to decentralized finance, you should now be equipped to avoid the risks and understand more about the newest advancements in this exciting and developing sub-industry.

LEAVE A REVIEW

As Independent Publishers it can often be difficult to get reviews.

Therefore, if you have gotten any value from
this book I would appreciate an honest review.

Many thanks.

Made in the USA
Las Vegas, NV
05 March 2022